THE SELFISH GENE POOL

An Evolutionarily Stable System

D. M. Wonderly

University Press of America, Inc.
Lanham • New York • London

Copyright © 1996 by
University Press of America,® Inc.
4720 Boston Way
Lanham, Maryland 20706

3 Henrietta Street
London, WC2E 8LU England

All rights reserved
Printed in the United States of America
British Cataloging in Publication Information Available

Library of Congress Cataloging-in-Publication Data

Wonderly, D. M. (Donald Mackay)
The selfish gene pool : an evolutionary stable system / D. M. Wonderly
p. cm.
1. Altruism. 2. Helping behavior. 3. Egoism. 4. Self-interest. 5.
Motivation (Psychology) I. Title.
BF637.H4W66 1996 171'.8 --dc20 96-20262 CIP

ISBN 0-7618-0382-3 (cloth: alk. ppr.)
ISBN 0-7618-0383-1 (pbk: alk. ppr.)

∞™ The paper used in this publication meets the minimum
requirements of American National Standard for information
Sciences—Permanence of Paper for Printed Library Materials,
ANSI Z39.48—1984

The Selfish Gene Pool

An Evolutionarily Stable System

Contents

Title page .... *i*

Copyright .... *ii*

Contents .... *iii*

Figures .... *ix*

Preface xi
Preface Notes xv

Acknowledgments .... *xvii*

Introduction .... *xix*
Egoism Run Amok xxii
    The political process xxii
    Social institutions xxiv
A Moral Alternative xxv

The Selfish Gene Pool

The Motivational Process xxviii
Introductory Notes xxix

1. The Issue 1
   Interpretations of Morality 2
   Altruism 6
      Intuitionism 8
      Provincialism 10
   The Good and the Natural 11
   The Emergence of Moral Behavior 13
      The biological aspect 14
      Are humans basically "good" or "evil"? 17
   Altruism and Social Convention 18
   The Dilemma 20
   Summary 21
   Chapter 1 Notes 23

2. Conflicting Views 27
   Sociobiology: From Genetics to Genaddicts 28
      Sociobiology and the self 31
   Culturalism—Biotheology or Theobiology? 32
   Psychological Interpretations 37
   Theological or Creationist Convictions 39
   Psychosocial Models 41
      The unit of selection 44
        genes and individuals 44
        groups 45
   Summary 47
   Chapter 2 Notes 49

3. Altruism and the Motivational Process I
   Existential Principles 57
   Evolution 58
   Adaptation 61
   The Holon 63
   Directional Holons—The Phenomenon of Life 65
   Systems 67
   The Holarchy and "Independent" Holons 70

Contents

Knowledge	71
Genes	72
Replicators, interactors, and lineage	73
DNA/RNA/protein	74
The Role of the Gene in Altruistic Behavior	76
Summary	78
Chapter 3 Notes	81

4. Altruism and the Motivational Process II

Behavior and the Role of Mind	89
The Function of the Sensory System	89
Mind	91
Cognition/affect	93
Belief	94
Desire	94
Classes of desire	95
Desire as instinct	98
Needs	98
Evaluative procedures	99
Emotion	101
Motivation	102
The self	103
Identification	104
Responsibility	105
Deliberation and decisions	106
Behavior	107
Behavior Adjustment Paradigm	109
Willing	111
Critical Behavior Characteristics	112
Reason, Morality, and Altruistic Behavior	115
Summary	118
Chapter 4 Notes	121

5.
The Sociobiological Thesis	127
The Sociobiological Argument	130
Kin selection or inclusive fitness	130
Reciprocal altruism	135

The Selfish Gene Pool

An Evolutionarily Stable Strategy	138
The prisoner's dilemma	139
Rebuttal	141
Ambiguity	141
Hamilton	141
Trivers	143
Dawkins	146
Barash	149
Mayr, Wilson, et al	149
Recognition alleles	150
Family dissidence	151
Friendship	152
A phantom affect?	153
The evolutionarily stable "strategy"	154
Research interpretation	158
A paradigm shift	159
The critical motivational flaw	161
The holarchic explication	162
Encouraging altruism!	163
Summary	166
Chapter 5 Notes	167
6. The "Selfish" Gene Pool	**177**
Species and Gene Pools	177
The Evidence	180
Altruism and the law	184
Legal ethics	186
Medical ethics	187
Thought experiments	188
The self-assertive desire	188
The self-protective desire	191
The self-transcendent desire	191
Altruistic behavior and moral credit	196
An altruistic gene?	197
Social, Political, and Educational Goals	200
A Recapitulation	203
Summary	206
Chapter 6 Notes	209

Contents

Glossary 215

References 219

Index 231

Figures

4.1 The Behavior/Adjustment Paradigm 109

6.1 Smoking vs. Punishment 189

6.2 Escape vs. Injury 192

6.3 Altruism vs. Egoism 194

Preface

The vast majority of scholars have concluded that the twentieth century has seen the most dramatic improvement in almost every form of civilized life since the beginning of time. It would be difficult to disagree with such an analysis. Unfortunately, such progress has been accompanied by a number of pronouncements and programs that have contributed to the moral and social decay of individuals, families, an nations. Three of such anomalies have probably been most responsible for these aberrations, one of which shall be challenged in this text. Each ostensibly represented a "scientific" breakthrough in the understanding of human behavior, and each proposed an interpretation that makes every human being a victim of uncontrollable forces, both internal and external.

The first of these programs was foisted on an unsuspecting public during the first third of the century by an inconsequential—though intellectually superior—Austrian physician, who remarked quite openly that he resented not being more highly appreciated by his colleagues, and that he planned to show them (and the rest of the world) how brilliant he was. Sigmund Freud thus commenced to create an edifice based on the contention that all behavior is controlled by unconscious motives, responding to the demands of the tyrannical "id."[1]

> In describing the id, Freud pained a picture of irrational, illogical desire which operates on the basis of a balance between desire for pleasure

and desire for minimal stimulation.... Freud viewed the primitive personality as a boiling kettle with steam attempting to escape at every joint. Life, he believed, begins with unbridled lust, immediate demand, and insane desire.[2]

Such an interpretation (while there may be much to recommend it as a therapeutic technique) to the extent that is accurate, frees everyone from responsibility. The tortuous explanations are, in fact, in many cases bizarre.

The elements of the unconscious system are defined in so many ways that a nightmare of confusion results. Such terms as instinct, longing, liking, drive, wish, need, desire, urge, motivation, emotion, impulse, and intention are interchangeably employed throughout the literature with little attention to their essential differences. Weiss, a prominent analyst, used the first ten of the above terms on one page to describe the same concept as well as (on the same page) such combinations as "instinctive drives" and "emotional motivation."[3]

In spite of—or perhaps because of—such esoteric phraseology, many psychologists have been enraptured by psychoanalytic explanations of behavior. The oedipus complex, which professes to explain incest, is seen as an ineluctable *inherited* concept. (Freud was an avowed Lamarckian in the belief in the inheritance of acquired characteristics), and thus not amenable to deliberate attempts at modification. Freud claimed, for example that "the basic underlying motivation behind Leonardo [da Vinci's] tremendous creative talent was a desire to sleep with his mother."[4] How convincing a thesis! Most absurd was the oxymoronic positing of a dynamic unconscious which houses desires, emotions, and similar affects that are meaningless except as related to conscious experience.[5]

The banner was picked up by B. F. Skinner and behaviorist psychologists during the second third of the century with the contention that individuals should not receive credit (or blame) for their behavior. "We are not inclined," Skinner said, " to give a person credit for achievements which are in fact due to forces over which he has no control."[6] While challenging Freud's notion of a "dynamic unconscious," Skinner agreed that people are essentially automatons, acting on cues from a variety of sources, all of which either reside outside themselves in the form of envi-

Preface

ronmental pressures or are triggered by instinctual forces. "There are no doubt minor instances of the struggle to be free...[but] we do not attribute them to any love of freedom; they are simply forms of behavior which have proved useful in reducing various threats to the individual."[7]

As in Freud's analysis, humans are described as *victims*, this time to the vagaries of internal and environmental forces. If an action is rewarded goes the mantra, the individual will learn to respond in a similar way at the next occasion on which that reward is offered. In this text, an interpretation of responsibility which challenges the behaviorist view shall be presented.

The third member of the triumvirate, the "selfish gene," which came on board as an inevitable consequence of the work of Crick and Watson on the nature of the genetic structure, is the subject of this book. Under that scenario the self serving gene rather than the individual is assumed to be the source and director of behavior. As a result, genuinely altruistic behavior is believed to be impossible. Individuals are driven by the pressure of inherently egoistic genes. They are *victims* of genetic influence. As for morality, Weiss proposed that "it is not...that we desist from aggression because we have a rigorous moral ideal, but rather, we have a rigorous moral ideal just because...we have renounced aggression."[8] Thus, morality, if it exists at all, is an outcome, not a cause of behavior. Dawkins was most extreme, proposing that we must "treat the individual as a selfish machine, programmed to do whatever is best for his genes as a whole."[9]

The doctrine has, of course, been excoriated by many philosophers, psychologists, and other professionals. Hoffman, for example, made the point that:

> The doctrinaire view in psychology has long been that altruism can ultimately be explained in terms of egoistic, self-serving motives.... It is always possible, when viewing an example of human action that appears to be motivated by an interest in the welfare of others to induce a hidden, unconscious or tacit self-regarding motive (e.g., social approval, self esteem) as constituting the real source of such behavior.[10]

Others are equally adamant. The claims of sociobiologists are ridiculed as absurd principles of human behavior—similar to the Freudian claim that scientists may ultimately discover a little creature or entity in the mind that represents the conscience. ("It would not surprise us if we were to find a special institution in the mind which performs the task of seeing that narcissistic gratification is secured from the ego-ideal.")[11]

> What the sociobiologist has done is to attempt to rid biology of the "mystery" of self sacrificing behavior and the notion of purpose; to get at what *really* causes behavior. Not some nebulous species centered genetic idea, but a real, finite little mechanism; a clever little entity that directs individuals, without their knowing why, to prefer their own kind; a molecular collection that reads self enhancing potential in the future behavior of a recipient of self-sacrifice.[12]

Unfortunately, the sociobiological analysis of behavior ignores the influence of a moral sense, and the fact that altruistic behavior (which they deny exists) is directed not toward the resident gene, *but toward the totality of the relevant gene pool*. In spite of the criticism that has been leveled at sociobiologist extremists, nothing has appeared in print that represents an alternative explanation for altruistic behavior. Most authors (e. g., Midgley) have focussed on the "hellish" nature of such a philosophical thesis. They are offended by the inhumanity that is inherent in sociobiological assumptions.

It is not sufficient to merely decry the proposal by respected professionals that greed is the norm; that our selfish genes are the villains. It is necessary to provide evidence that their thesis is inaccurate. In this text, an explanation that encompasses the moral sense as a critical factor, and legitimizes the existence of genuinely altruistic behavior will be presented.

Preface

Preface Notes

1. Brill, A. (1938), *The basic writings of Sigmund Freud.* NY: The Modern Library
2. Wonderly (1991), p. 238
3. *Ibid.* p. 329
4. Freud, S. referenced in Levine (1975), p. 4
5. The problem with such perversion of language terms will be discussed in Chapter 4.
6. Skinner, B. (1971), p. 44
7. *Ibid.* p. 26
8. Weiss (1960), p. 339
9. Dawkins (1976), p. 71
10. Hoffman (1981a), p.41
11. Freud, S. referenced in Riviere (1959), p. 82
12. Wonderly (1991), p. 249

Acknowledgments

The material presented in this text is based in considerable part on a series of conferences with one-time students and "fellow travelers" with the PSI (Prevention: Systems Intervention) organization held during the past two years. It grew out of the conviction that the public is, once again, being duped by a so-called "scientific" approach to a significant problem—that of the nature and role of altruistic behavior.

Participants included Dr. Steven Rosenberg, current president of PSI, Dr. Joel Kupfersmid, itinerant author and provocative thinker, Dr. Kathleen McNamara, now a Cleveland State University professor, and Dr. Linda Carter, our resident "psychic" and creative analyst. Each of these individuals seemed to have a grasp on separate aspects of the problem. Meetings were exciting as well as productive. Positions were proposed and challenged. What appears here represents the resolution of many such concerns, hopefully to the satisfaction of those who were involved.

The final preparation of this text owes a great deal to the patience, and skill of another colleague Laura Guerreiro Ramos. It is quite a marvel of esoteric expertise that her performance as a compositor, editor, secretary, artist, and dedicated professional is always underplayed.

The Selfish Gene Pool

An Evolutionarily Stable Strategy

Introduction

Flowing beneath every human action is the current of ethical significance, and in all ages and places, questions about moral conduct and moral principles are posed and answers attempted.

Albert & Denise

In 1979, George Wald, a highly respected biologist confessed that the immense magnitude of problems associated with the theory of evolution led him to concede "that the spontaneous generation of a living organism is impossible. Yet here we are—as a result, I believe, of spontaneous evolution."[1] A similar situation exists in the search for the genesis of morality and its expression in altruistic behavior. An analysis of the reproductive process seems to lead to the conclusion that truly altruistic behavior is impossible. Yet here we are—experiencing empathic states which, if followed to their limits, ought to result in the genetic contribution of altruistic individuals dropping out of the gene pool; their fitness, inclusive or otherwise, being inadequate to the task of achieving representation in subsequent generations.

The Selfish Gene Pool

For centuries prior to the coming of age of biological research, philosophers struggled with the question of whether God was the originator of, or is subject to, the demands of moral law. In some circles, that argument is still pursued. In other areas, however, the concerns of theologians have been supplanted by those of molecular biologists, geneticists, ethologists, and sundry behavioral scientists that have come to be known as *sociobiologists*, or more recently as *evolutionary biologists*. It is their contention that social behavioral characteristics are as genetically prescribed as are the physical. Today the focus is on the issue of whether ostensibly altruistic behavior can be thoroughly accounted for by genetic determinants, or whether the basis for ethics lies in the fact that humans are, unlike their animal forbears, "self-perfectible" beings; capable of rising above their naked, determinist, genetic inheritance.

The problem has persisted into the last decade of the twentieth century as a source of concern in biological circles. Mayr points out that there is a "seeming irreconcilability of opposing opinions,"[2] adding that: "We are still far from a resolution of questions surrounding the role of genetics in human ethics."[3] At issue, is a wide variety of social and personal factors that may have a significant bearing on educational, as well as political programs.

One cannot help being impressed by the insistence of biologists that scientific accuracy demands the acceptance of a fact that nature puts before us. The "fact" to which they refer is the control exerted by genetic inheritance over everything from hair color to sexual preference. The concern of this text is with the contention that the demand of reproductive fitness mandates that the welfare of each (gene transporting) individual and perhaps immediate family members is the focus of *all* human behavior. The diametrically opposed perspective—that genuine altruism is a characteristic of much human activity—shall be supported through rational analysis as well as with extensive behavioral evidence.

Dawkins tossed a small incendiary into the arena of biological thought with his publication of *The Selfish Gene* (1976), and *The Extended Phenotype* (1982), in which he specifically denied that more inclusive entities ever take priority—that outside the domain of the nuclear family, all behavior is motivated by the urge to maximize the reproductive potential of the behaver. "Much as we might wish to believe otherwise," he said, "universal love and the welfare of the species as a whole are concepts which simply do not make evolutionary sense."[4] And Barash argued that:

Introduction

"True altruism implies that the altruist's inclusive fitness is actually reduced. [It] should occur only adventitiously, if at all."[5]

Biologists have aligned themselves with one side or the other of Dawkins' claim. (Many sociobiologists, such as Dawkins, Trivers, and Hamilton have themselves, in fact, made statements supporting both sides of the argument.) By the 1990's the literature was suffused with books and articles defending or attacking the premise that all behavior is geared to the interest of the behaving individual. Wilson, who has spoken at various times on both sides of the issue, contended at one point that since genes share a common ancestry, there is a "propensity of altruism [to] spread through the gene pool,"[6] as is proposed in this text. Such a view is, however, rejected by Dawkins and his fellow sociobiologists.

Hamilton introduced a theory of "kin-selection" which proposes that relatives are assisted in proportion to their genetic distance from the behaver. Trivers developed the principle of "reciprocal altruism" which is based on the tenet that ostensibly altruistic behavior is performed in the anticipation of future recompense. For all such theses, the underlying principle is that unless individuals acted "selfishly," their genetic material would disappear from the ensuing gene pool.

This extreme view is countered by many anthropologists, psychologists, psychiatrists, and other social scientists in what in many instances seems a desperate attempt to save humanity from the terrible prison of biological determinism. Their contention is that human life transcends that of simpler organisms and cannot be understood through the most thorough analysis of the behavior of animals of any type. They refuse to concede that moral judgments and altruistic behavior are based solely on efforts to further the personal interests of each individual to the exclusion of those of others.

This text provides a refutation of the arguments and the evidence offered to support the "selfish gene" hypothesis, particularly in its most radical form. The subject matter will be the nature of morality and the role of altruistic behavior. The purpose is to analyze the impact of the acceptance of one or another point of view on the type of educational and political programming that is to be recommended for the 21st century. Equally importantly, it should provide a basis for the evaluation of professional ethical codes, many of which, like those that guide the behavior of elected officials have been interpreted in such a manner as to reflect the acceptance of a self-serving philosophy; of a purely egoistic view of life.

Egoism Run Amok

The political process

Toward the end of the eighteenth century a group of American colonists met for the purpose of forging an instrument designed to make possible the creation of a nation which would provide a maximum of freedom to its citizens, while maintaining respect for a (Judeo-Christian) God. Although the rights of the individual were seen as a defining principle, respect for the welfare of others was expected to be taken into account. Decisions were based on a moral philosophy which included the acceptance of an altruistic urge as a significant human characteristic. Over time, the nation that they founded became a model of positive human enterprise. The Statue of Liberty became a shrine; a beacon that declared to the world that human decency had found a place of residence. America was seen as a melting pot of people of disparate backgrounds with common interests who shared a willingness to accept a diversity of life styles, opinions, and convictions.

In spite of the fact that many of the founding fathers were something less than paragons of personal virtue, the principles of equity, of mutual responsibility, of the acceptance of curbs on many forms of behavior, were accepted as morally, as well as legally, binding. An infrastructure was created that made the American political system responsible to the precepts of "laws" rather than "men." Right action was believed to be recognizable, though the particulars in cases of litigation may vary. Such practices as that of Sir Frances Bacon, Lord Chancellor of England ("I usually accept bribes from both sides so that tainted money can never influence my decision.")[7] were decried.

At the end of the nineteenth century. Americans spoke from a position that they sincerely believed to be one of moral superiority. Their strength was based on a concern for the welfare of individuals (in spite of the fact that they had not yet recognized that minorities and women are fully human). The situation has, however, altered drastically. At the close of the twentieth century people have come to reject America's moral leadership. America is considered by many to be ethically moribund. Americans are seen as takers, exploiters, the worst examples of what wealth and associated greed can produce. There is serious reason to question whether, given the prevailing state of affairs, this nation shall survive the moral pitfalls associated with political and other forms of social depravity.

Introduction

The acceptance of sociobiological principles raises serious questions. How can people be criticized for doing "what comes naturally?" For "looking out for number one?" For "feathering one's own nest?" For deceit, dishonesty, blatant hypocrisy, and worse? In the final analysis, one need only agree with the contention that moral sentiments are no more than an instance of the irrelevant emotional experiences that behaviorist psychologists have proclaimed them to be. One has only to buy into a biological philosophy that identifies self-serving behavior as the goal of *all* human motivation—an ineluctable characteristic of behavior.

Although the two philosophies developed independently of each other, the biological principle that all people are inherently selfish, and the political explanation that disgraceful behavior is legitimized by the fact that "everybody does it," complement each other very conveniently. Congressmen who exempt themselves from laws they write for the populace, who have created lifetime sinecures for themselves, who are bought and sold by political PACs, whose personal lives are often a series of debaucheries, excuse it all on the contention that such behavior is human; normal; to be expected. Persons expelled from public office become wealthy on the basis for the demand for their services in the private sector. Dishonored ex-presidents become "elder statesmen." Ashman complained that:

> American justice is choking on judicial pollution.... It is no longer a question of occasional corruption, but a decided pattern of conflicts of interest, chronic bribery, profound abuse of office, nepotism, infamous sexual perversions and pernicious payoffs.[8]

Such behavior pervades the political landscape A president is elected who has a history of flagrant marital infidelity, avoiding his military service obligation, and practicing questionable ethical behavior as a governor. He boasts that in winning the election he has "overcome the challenges of party campaigning," (for which he apparently believes he should be rewarded). "Character" is considered less important than the dissemination of promises, empty or not. An individual who exudes charisma, and "looks presidential" appeals to a plurality of voters. Although such a person is recognized by most Americans as being reprehensible, supporters contend that his behavior should be condoned, on the argument that no one is perfect. Standards of decency are abandoned in favor of "compassion" and "understanding."

A first lady whose behavior may land an ordinary citizen in jail is considered non-assailable. Royalty, rejected in the eighteenth century, is reinstated at the close of the twentieth. But such activity is rampant among elected officials, perhaps reaching a disgraceful crescendo in the behavior of many Supreme Court members who operate as if they were competing in a lifelong popularity contest; persistently rewriting the constitution to provide support for a government that caters to the greediest among us.

Perhaps Feder was right in his contention that "since the beginning of our species, [an] innate 'predisposition,' however beneficial or necessary, has delivered us into the most grisly slaughters, into the hands of the most unconscionable manipulators;"[9] into the hands of those who hold sufficient political power to pervert the American Dream, and the acumen to realize that the vast majority of people have neither the force nor the inclination to resist.

Social institutions

What is taking place in the political sphere is reflected in every organization that has been created for the purpose of civilizing the American people. In the education arena, from kindergarten through postgraduate university settings, teachers have forgotten their obligation to serve. Students are "used" to serve professorial interests. Families have turned child rearing over to television sitcoms and Hollywood extravaganzas, where the common theme is to *take*, to *use*, to *embarrass*, to *destroy*. Notions of giving, of sharing, of accepting responsibility, are conspicuously absent. Churches are castigated for their emphasis on what are labeled "radical" approaches to the problem of moral decadence. Selfishness has, in fact, become the norm that sociobiologists proclaim. We seem hell bent on proving that Dawkins, Trivers, and their colleagues who insist that greed is the basis for all human behavior, are correct.

One of the results of the polarization that self-orientation has spawned is that America has become a nation of subgroups of the "victims," that Freud, Skinner, and associated pessimists have described; ready to litigate at the drop of a hat. The issue has been described by many journalists. Feder points out that:

> "Like a printing press in the treasury of a banana republic, the rights industry works overtime. We are now into rights hyperinflation, the

Introduction

minting of new entitlements at such a dizzying pace that it threatens to thoroughly debase a noble concept."[10]

Such an attitude has resulted from a Sodomlike response to the many freedoms that have been developed in a democratic society. Attempts to control it are unlikely to succeed if it is assumed that a rational approach is appropriate. "Rights inflation," says Feder, "is fueled by greed, belligerence, envy, and a desperate need to escape personal responsibility. Persons in the grip of these emotions are rarely open to reason."[11] And who are these people that insist on being catered to, cared for, continuously nurtured by the political social system? O'Rourke, in his text *The Parliament of Whores* states it clearly. Take *his*. Give *me*. "The trouble is, in a democracy the whores are us."[12] If sociobiologists are right, there is no need for anyone to respond to moral demands. If they are not, there ought to be ways to stimulate the urge to build a society that recognizes the value of altruistic behavior.

A Moral Alternative

There are a variety of worthwhile goals involved in the development of individuals who accept the responsibility inherent in effective personal and social relationships. Guisinger & Blatt propose that: "The recognition of the importance of both self development and interpersonal relatedness...can provide a theoretical basis for appreciating and encouraging the development of these essential dimensions in all members of society."[13] It is, of course, equally important that such a "theoretical basis" be translated into practical programs, which requires the appreciation of the fact that effective interpersonal relationships call for the acceptance of a moral sense that influences altruistic behavior, and a denial of the claim that all behavior is based on self-serving motives.

Beyond the fact that the wide practice of altruistic behavior would be advantageous to a community, is the equally important consideration that through risk and sacrifice *people learn to love*. The pertinent principle is that individuals love most deeply those to whom they give. In practicing altruism people learn to identify with those that they assist. An expanding self emerges. The individual who receives assistance may develop a feeling of appreciation and respect, but love is an affect that develops out of a sense of commitment; of involvement; of obligation.

And how may the behavior of those who do not—or cannot—love be interpreted? James Wilson, in his book, *The Moral Sense,* proposed that, "the naughty and the selfish are just one case of moral inadequacy. The other, or one other case is presented by the indifferent and the despairing. Both are examples of inability to love, which is why the notion of love turns out to be central."[14] He argued that loving provides a kind of safety, that it is "a matter of *having a world to live in,*"[15] and that "safety consists precisely in the rejection of autism, in being able to come to terms with and accept the world as it is."[16] That philosophy; the equating of loving with security, with peace, and the acceptance of others leads to profit to the individual—an apparently "selfish" outcome. However, the basis for the feeling of moral satisfaction is in what one does for others; what altruistic behavior is undertaken.

The effort to initiate such a life style may seem difficult. Wilson once again offers a suggestion:

> We may be convinced that it is impossible for us, even with our best endeavors, really to love another person, or enjoy a job of work, or even feel at peace in our own home.... [But] once we have taken the first steps in the quest for love, the quest itself becomes interesting, exciting and lovable. We learn to enjoy becoming good, as well as being good.[17]

Such terms as responsibility, obligation, and commitment, as well as the altruism they engender are essential to such an enterprise. They are meaningful to the extent that they bind individuals to more comprehensive entities. An individual who experiences a sense of responsibility—and thus of love—expresses that feeling in behavior directed toward particular groups; toward organizations or communities which the individual invests with value.

Values can, of course, be taught if one means by that the teaching of who or what are appropriate meritorious objects, as when a mother teaches her children to respect a teacher, or a community cleanliness code. On the other hand, in many instances individuals find reason to respect institutions, or other individuals without formal training. In all such instances the notion of *community*, described earlier, is involved; where social relationships are considered significant; where commonalities are recognized between *self* and *other*.[18]

Introduction

It is because of the perplexity of such questions that the concepts of ethics and morality become confused. It is why many theorists, in their failure to find universal ethical principles, deny the legitimacy of a moral sense. The confusion can be addressed however, by appreciating that altruistic behavior is considered by civilized people to be *universally* good. It is born of an innate sense of responsibility that is satisfied by the practice of "good works," and the loving feeling that such behavior creates.

The opportunity to share should be encouraged at an early age. Unfortunately, the approach to developing altruistic behavior in children is complicated by the nature of the social system. Wilson enunciated the problem. "Young snakes emerge from their eggs and slide off into life noncontroversially.... The growing child has to negotiate his own feelings and desires and emotions and thoughts *for himself*.... [He] has to give up his original objects of desire because they are forbidden or incapable of attainment.... [However], collectively, we can best operate through education; more is gained...by trying to ensure that our children are taught to love and to reason properly about morality than by trying to change ourselves as adults."[19]

People must be encouraged to seek to assuage the urge to give and to share, just as they are encouraged to study, which may be considered a cost related to learning to read. They must learn that escaping into a world created by indulging in the use of drugs of various types; of entering into an underworld euphemistically referred as a "drug culture" is self-defeating; destructive; ultimately often deadly. It is difficult under the best of circumstances to get the attention of a significant number of people, especially when the behaviors that educators, politicians, parents, clergymen and others are recommending requires that indulging in behaviors related to self-oriented desires must be sacrificed.

The struggle to maintain support for altruism is made more onerous by the continual assault on the principle by the exhortations of members of a liberal press—a coterie of individuals who are steeped in a philosophy of self- indulgence, and moral relativism. Greed and avarice are the normal condition, they say. A well known columnist, in defending the scurrilous behavior of a United States president said: "Lucky for him he was not elected as a role model or a moral authority."[20] But not so lucky for the millions of Americans who are taught through such articles that selfishness, dishonesty, and similar behavior cannot be avoided. Professionals and politicians cannot escape their responsibility to others by arguing that selfish behavior is inevitable; that immoral and unethical behavior is to be excused because "everyone is doing it." Any hope for a positive change in the moral direction that America is going will require a fresh look at the arguments being offered for understanding (and often for exculpating) those who practice selfish behavior consistently.

The Motivational Process

The goal of this text is to make the case that the "reason" referred to by Feder and others is the wrong evaluative technique to which an appeal must be made. The reasonable man is *reciprocally* altruistic. The moral man is *genuinely* altruistic. There is a critical role for altruistic, as well as other types of moral behavior, and human beings can learn to recognize obligation to others as a legitimate canon. Furthermore, governmental and other agencies that deal with morally relevant behavior have a responsibility that exceeds that of private citizens. This is not to deny the importance of the pursuit of self-interest, and even that such behavior must often take precedence over the care of others. However, altruistic behaviors are based on inherent affective experiences, and the desire to act to enhance the welfare of others is an urge as compelling as any other.

It is conceded at the outset that if biological self-interest is the goal of all behavior, no deliberate action could be taken that furthered the interests of any other individual except as a means to an end benefiting the behaver. However, the position shall be developed here that as the intellect developed, a countervailing force—a moral sense—emerged that has the effect of maintaining the fitness of groups. A model will be proposed that identifies gene pools as *evolutionarily stable systems*, in the interest of which altruistic behavior has evolved.

In this, as in similar endeavors, conclusions are limited by the extent to which persuasive arguments can be mounted. It will be perhaps hundreds of years before an understanding of the genetic process is sufficiently complete for a definitive explanation to be provided. Meanwhile, sociopolitical decisions must be made that have a significant impact on the way in which humans interact. It is critical, therefore, that every possible interpretation of putatively altruistic behavior be examined. The model presented here represents an analysis of behavior and the motivational process which provides an interpretation of altruism that is consistent with observed human (as well as many animal) activities.

Introduction

Introductory Notes

1. Wald (1979), p. 48
2. Mayr (1988), p. 6
3. *Ibid.*
4. Dawkins (1976), p. 2-3
5. Barash (1977), p. 97
6. Wilson (1975), p. 3
7. Bacon quoted in Ashman (1973), p. 13
8. Ashman (1973), p. 3
9. Feder (1993), p. 21
10. *Ibid.* p. 311.
11. *Ibid.*
12. O'Rourke (1991), p. 233
13. Guisinger & Blatt (1994), p. 110
14. Wilson (1987), p. 57
15. *Ibid.* p. 63
16. *Ibid.* p. 64
17. *Ibid.* p. 125
18. But that difference is often only a matter of focus. Commitment is time, place, and situation specific. Where, then, does altruistic behavior become appropriate? If "charity begins at home" where (and when) is home? Or am I equally responsible for the welfare of all people—and perhaps, to a lesser extent perhaps to all forms of life?
19. Wilson (1987), p. 125
20. McGrory (1994), p. A-9

Chapter One

The Issue

> *Why do we have empathic feelings? Why do we applaud philanthropic behavior and actions that follow rules established by social, religious, and legal custom? Why do we feel respect and admiration for some behaviors; resentment and revulsion for others? Is it no more than a contingency of our DNA molecular structure? The expression of ideals established by an omnipotent maker? Or are altruistic behaviors perhaps no more than conventions, designed to emphasize the benefits of social interaction, and to enhance the quality of human life?*

From the time that the helical nature of the DNA macromolecule was discovered, every branch of science and every school of philosophy has focused attention on the role of the gene, and the primacy of the individual. Behavior is believed to be based on rational rather than moral principles and fitness needs demand that selfishness be recognized as the sole basis of human interaction. Genuine altruism is said to be non-existent. The history of such a view extends back as far as Socrates, who said that "self-seeking, in some sense, and more particularly self-preservation, govern all human behavior."[1] Spinoza provided a rationale for the denial of the requirement of appealing to a moral sense, proposing that "men who are governed by reason...desire nothing for themselves which they do not desire for other men, and that, therefore, they are just, faithful, and honorable."[2] An analysis of Spinoza's position would reveal either that such individuals are foolish, (since they risk the loss of personal profit) or assume that such behavior will ultimately result in advantage to themselves. They could not be guided by a rational sense that spontaneously put the welfare of others above their own. That conviction, now supported by a powerful biological principle, has captured the allegiance of geneticists, molecular biologists, psychologists, and anthropologists. It suffers, however, from one serious shortcoming. *It is wrong.*

Interpretations of Morality

Before embarking on a study of the nature and significance of altruistic behavior as a manifestation of a moral sense it may be prudent to specify the nature of the concepts that are being studied. A critical question is whether the focus should be on *attitudes* toward beneficent behavior, the *motives* for such action, the *behaviors* themselves, or the outcome or *consequences* of behavior; whether behavior should be evaluated in terms of conformity with established norms, or by the extent to which individuals perform in accordance with personal criteria of propriety.

Consider first the general terms "morality" and "ethics." Definitions provided by philosophers are not particularly enlightening. Flew, using the terms ethics and morality as synonymous, proposes that they are related to the determination of the meaning or purpose of life, saying that "once this purpose has been clearly established, any moral principle or virtue can be assessed in terms of the contribution it makes, or possibly fails to make, toward this end."[3] But how shall the purpose of life be determined? And can any such standard have objective characteristics? Flew's definition would seem to provide a basis for the development of a situational ethic.

Runes distinguishes between ethics as the study of values (what *ends* ought to be pursued), and as a source of obligation (what *behaviors* should be undertaken to achieve those ends).[4] The former interpretation may provide an excuse for manifestly undesirable behaviors—with ends justifying means, a teleological approach—while the latter harnesses action to outcome, perhaps supporting a form of *rule* utilitarianism.[5] While Flew says that the layman's view of ethics "suggests a set of standards by which a particular group or community decides to regulate its behavior,"[6] Runes contends that although morals and ethics are sometimes equated, "more frequently [morals] are used to designate the codes, conduct, and customs of individuals or of groups."[7] That is, morality is defined in terms of motives. A quite improper interpretation.

Kohlberg offered a psychological definition, denoting "the most inclusive conception of the good and the right as the ethical."[8] Once again, no distinction between ethics and morality is drawn. Albert and Denise did little to clarify the issue, saying that "ethical theorizing is concerned with the construction of a rational system of moral principles and...with the direct and systematic examination of the underlying assumptions of morality."[9] Here, not only are ethics and morality intertwined in a confusing manner, but reason ("rational system") is brought into the arena, which will be shown to represent an insupportable thesis. Waddington offered an explanation which, once again, was of little value in discriminating between the concepts. Man, he said, has a "built in predisposition towards certain ethical values which have the same general relevance to human society as do the Euclidean axioms of geometry to the material world."[10] Such interpretations are not only conflicting, they seem to suggest that almost any definition of the terms morality and ethics is appropriate, which is apt to be quite misleading. A critical distinction must be drawn between these concepts.

Morality will, for the purposes of this text, be restricted to referencing an *innate sense* by which the propriety of behaviors is established, while ethics shall refer to codes of conduct that are established on the basis of that sense. Propriety is, of course, considered by many to represent no more than conformity to social expectations. The focus here, however, will be on the *source* of such expectations; the *genesis* of feelings of obligation; the *origin* of the sense of responsibility.

Ethical codes prescribe socially acceptable behaviors. They mandate some activities, while proscribing others. Common usage will make it obligatory that the terms be used interchangeably in many instances. However, it must be clear that they are distinct concepts, with morality referring to a feeling state, and ethics to a set of behaviors assumed to be generated by that feeling. The former is an affective experience, the latter represents the application of rules based on the influence of that sense.[11] Ethical behaviors are learned. The moral sense is inherent. It is essential that scrupulous attention be paid to the maxim that "*one learns what is 'good' [i.e., ethical behavior] but has innate knowledge of what 'good' [morality] is.*"[12] It is perhaps more important to recognize that ethical behavior is a form of altruism, as shall be developed.

When the promulgations of moral theorists (e.g., Aristotle, Spinoza, Kant, Bentham) are studied, the subject matter is discerned to be ordinarily rather that of ethics, as the term is defined here, and they are best described as *normative* ethicists. Many of the behaviors discussed in this text represent examples of normative ethics, or beliefs regarding which particular behaviors, or classes of behavior, should or should not be undertaken—assumptions and statements that are evaluative. However, the ultimate concern will be with the field of metaethics, which deals with the meaning, or justification of normative ethical philosophies.

Hancock (1974) argued that it is difficult to distinguish clearly between normative and metaethical issues, accepting Steven's (1944) distinction that metaethics is simply involved in the business of clarifying the meaning of ethical terms. The concern here is rather with the origin of such terms, and how they have come to stand in need of clarification. The approach taken in this text will have characteristics of both *cognitive naturalism*, in the contention that normative ethical principles can be derived from existential conditions, and *cognitive non-naturalism* because of the positing of a special (moral) sense. Furthermore, it shall be argued that moral principles (but not ethical behaviors) are universalizable, thus subscribing to the school of moral *objectivism*.

Consider Kant's "categorical imperative," Mill's "greatest happiness" principle, or Rawls' revival of the "social contract" philosophy. In each instance the recommended behavior is based on the acceptance of some implied axiom. Recommendations speak to action that is appropriate only if one assumes the influence of an unlearned moral postulate. Since such premises are ostensibly uncertifiable, the force of ethical proposals is debilitated. The palpable result of this tenuosity is manifested in the Sodom and Gomorrah of twentieth century civilization where moral standards are trashed and traditional behavioral codes are held up to mockery.

Of most concern is the issue of how the moral sense came to evolve—if evolve it did—in the first place. The answer to that question should provide information essential to deciding what role the genetic determination of behavior plays in ordinary human social intercourse, what institutions are, or should (sic) be involved in the maintenance and encouragement of morally positive behaviors, and what steps may be taken to improve such institutions. In the process, the focus shall be on behavior that in many cases appears to violate reason, in that it risks the well-being of the behaving individual.

Consider the position taken by G. E. Moore, who argued that the term "good" is unanalyzable. "Good is good," he said, "and that is the end of the matter.... It cannot be defined, and that is all I have to say about it."[13] Moore was by no means contending that the notion of goodness is unimportant, but only that he believed it to be a simple concept, one that cannot be explained by any of the ordinary methods of defining a term. The term "good" as Moore employed it refers to propriety, and is synonymous with "moral." When applied to ethical behavior, it refers to the extent to which that action is based on, or is consistent with, some moral principle. Many philosophers have expressed discouragement with efforts to provide a definition of morality that can be generally supported. Hare, for example, said that it is probably best to "admit that the word [moral] is ambiguous...and to define a use for it which will mark out those uses of 'ought' and 'must' in which we are primarily interested."[14]

Warnock agreed, saying that "morality does, so to speak, shade off into other things, with a disputable area round it rather than a tidy frontier."[15] Prichard dealt with the nature of the good as Moore did but added, "for anyone to *ask* this question [what good is] it to imply that he already *knows* what particular actions are just."[16] Sharp agreed, adding that satisfaction would not accompany altruistic actions "if the desire for the realization of the other person's welfare had not been there in the first place."[17] Prichard's interpretation is not consistent with the argument of this text. Although a moral sense is intuitively recognized, the vast majority of human behaviors are neither instinctively nor intuitively known, as Prichard proposed. As to Sharp's contention, the *desire* for the welfare of others must be distinguished from the *moral evaluation* of such behavior.

Some moralists have proposed an interpretation of morality that is either vacuous (e.g., emotivism; the communication of one's feelings) or describes behaviors that are ultimately self-serving. Baier, for example, said: "It will be clear to everyone that universal obedience to certain rules overriding self-interest would produce a state of affairs which serves everyone's interest much better than the unaided pursuit of it [self-interest] in a state where everyone does the same."[18] This is an absurd argument. The abstract notion of what would be better for each person if everyone agreed to some form of behavior has to be balanced against the fact that any particular individual will be served best if everyone *except that person* follows such rules. Worse, it represents no form of morality at all, since the behavior that Baier proposes is, once again, designed to serve the interest of each behaving individual! Profit to others would be no more than a contingency.[19]

Many similarly defective arguments are offered that purport to show how individuals come to act morally. Markl proposed that altruistic behavior may come from "the ability to be aware of the consequences of one's behavior and from there to develop the concept of responsibility for one's actions."[20] While Markl referred to such a process as only a *requisite* for morality, it should be clear that there is simply no path from recognizing consequences to acting responsibly. Why is it reasonable to act responsibly unless one expects to be rewarded? Or unless the "self" is more inclusive than the biologic entity as shall be developed in this text.

Altruism

The determination of whether or not altruism is a characteristic of the actions of members of the animal kingdom and—most importantly—of human behavior is beset by many problems. Foremost among these is the issue of precisely what the term signifies. Runes defined it as "the cult of benevolence,"[21] suggesting that philosophers have seen it as representing "the pursuit of the good of others, whether motivated by either self-centered or other-centered interest, or by disinterested duty."[22] Brandon and Burian take the position that altruistic behavior "decreases the fitness of the actor while increasing the fitness of the recipient."[23] And Dawkins says "biologists define behavior as altruistic if it favors other individuals at the expense of the altruist himself."[24]

While definitions of this type may provide useful *descriptions,* or genetic configurations of altruistic behavior, they do not deal with the force that sets such behavior in motion. Furthermore, they do not address such questions as whether the behavior need be deliberate, what goal must be sought, or even whether moral credit is applicable to beneficent action—should individuals receive moral credit for acting altruistically?

Altruism shall be defined here as:

> the deliberate behavior of any individual, or group of individuals initiated by a *desire* to enhance the welfare of related or unrelated others, at potential or actual cost to the behaver, where no possible benefit accruing to those performing such actions can be discerned either immediately or over an extended period of time.

The experiencing of a desire is the operative factor. Individuals are programmed genetically to find altruistic activity gratifying.[25] The performance of an altruistic act is a manifestation of the acceptance of the risk involved. Ostensibly altruistic behavior is *evaluated* on the moral scale. This is a difficult but critical factor. The feeling that one wants to see the lot of others improved is entirely separate from the willingness to risk or sacrifice personal security in an attempt to achieve that end. What is *wanted* is the gratification of one class of desire; what is *done* is to risk one's assets. Propriety is based on the influence of the moral sense.

Sacrificial behavior may be focused on the result of the action itself or may be carried out in the interest of the possible consequences of the behavior. One may contribute to the welfare of others merely because of the positive feeling associated with the act, or may do so in contemplation of what ultimate benefit the recipient may derive from the benefaction. Under this interpretation the actions of many advanced forms of animal life may be said to manifest altruistic behavior. The broadest possible interpretation is applied here since it is proposed that all other-directed behavior is based on the same (moral) principle.

In dealing with such practices as respecting the rights of those of different racial, religious, and ethnic classes, protecting the weak and helpless, etc., it may seem that the concept of altruism as it is described by biologists is being distorted, in that sacrifice is not necessarily involved. However, such practices as sharing assets certainly represent a diminution of the power of the donor, albeit ever so slightly. That distinction shall be shown to explain behaviors from the philanthropic to the self-sacrificial. Although it is ordinarily thought of as behavior that puts one at risk in the interest of the welfare of others, altruism should be recognized as denoting a more general concept.[26]

The argument to be developed here, is that from the simple "excuse me" that accompanies a sneeze, through the sacrifice of time and money by philanthropists, to the sacrifice of one's life to preserve those of fellow humans, an altruistic propensity which is based on an *intuitively known* moral sense is manifested. No behavior deserves the appelation altruistic unless it is deliberately, and willingly performed. However, many of the actions of primitive organisms provide evidence that they are endowed with a propensity to act in ways that profit the gene pool at cost to the individual where there is no appreciation of the purpose of such behavior.

An analysis of behaviors catalogued in ethical codes reveals that most are—implicitly or explicitly—rooted in altruistic concerns. In each instance the obligations, rather than the rights or privileges of those who subscribe to a particular code are enumerated. They identify behaviors to be carried out or avoided in the interest of others, and they ordinarily provide sanctions for transgressions. Behavior based on social regulations manifests an unselfish concern for the welfare of others, and altruism shall be shown to be at least implicitly, involved. If those who contend that all behavior is ultimately self-serving are correct, there can be no justification for recommending altruistic behavior except as a mask for disguising selfish interests. If, on the other hand, humans experience a willingness to share, to accept obligation, to aid others, etc., though such a sense be expressed through learned behaviors, there is reason for optimism regarding the potential value of educational programs designed to foster the encouragement of other-directed impulses.

Intuitionism

The contention of this text—that the moral sentiment is based on some unique mental property—defines it as an example of moral intuitionism. Moral propriety is understood by these philosophers to be a matter of insight or apprehension. Reid pointed out that some philosophers "ascribe this [morality] to an original power or faculty in man, which they call the *moral sense,* the *moral faculty, conscience.*"[27] Huxley said that "the moral sense is a very complex affair—dependent...in part on an innate sense of beauty and ugliness."[28] For such thinkers, the moral sense provides an impetus which is separate from that which drives most behavior. In this text, it is defined as representing an evaluative technique.

The notion of "innate ideas," or the belief that the individual comes into the world possessed of some *a priori* knowledge goes back at least to Plato. Regarding morality, he said that "moral values are ideal, yet objective essences, apprehended intuitively at the end of a process of dialectic."[29] The significant reference is to his argument that values are recognized intuitively. But the concept of "ideas" in general has been contentious. Some, for example, have described innate knowledge as representing a set of "principles" or standards against which experiences are evaluated. For these philosophers, humans are assumed to be imbued with a moral sense, or capacity to recognize good and evil at first experience—not requiring any deliberative process to intervene.

Ruse presented the view of a variety of philosophers who support the notion of a moral faculty, and who argue that "morality, and our appreciation of it is something above and beyond the forces of natural selection."[30] And since evolution is progressive, "because the end product has more worth than the beginning crude life forms it is morally incumbent on us at least not to hinder the evolutionary process, and perhaps, even to cherish and positively aid it.[31] Hume took a similar position, arguing that in attempting to understand the vice inherent in a criminal act:

> You never can find it, till you turn your affection into your own breast....
> Here is a matter of fact; but it is the object of feeling, not of reason.
> [W]hen you pronounce any action or character to be vicious, you mean

nothing, but that from the constitution of your nature you have a feeling or sentiment of blame from the contemplation of it.[26]

The distinction between morality and reason is critical. At this point, however, the intuitionist interpretation proposed by Hume, Ruse, and others, is obvious. They conclude that moral evaluations are based on affective or feeling states. The position taken in this text is consistent with that espoused by both Hume and Ruse, with the reservation that the moral faculty, or moral sense is not, as Ruse proposes "above and beyond" the influence of genetic factors, but represents a genetically implanted evaluative mechanism.

The opposition came from the Locke/Mill school where the conviction that people are born intellectually—as well as morally—naked (the *tabula rasa* philosophy) attracted those who see empiricism and the preeminence of science as the road to knowledge of any kind. These philosophers, however, like the sociobiologists whose work will be analyzed, have failed to make a convincing case. In arguing that moral behavior is ultimately selfish, representing some form of social contract, empiricists obviously assume that other members of society will reward the waiving of personal interests. But on what basis should such a response be expected?

To assume that society is a rational "invention" either requires the acceptance of an innate (rational) sense, or assumes that one learns to be rational. The latter position is still embraced by a number of philosophers. Some argue that the moral nature of behaviors or events is learned through experiencing them. Laird, for example, said that "some writers hold...that our acquaintance with good and evil is akin rather to perception than to reasoning. We define the good, they say, in particular cases, and beyond that we cannot safely go."[27] This is, of course the position taken by Moore. Such theories have been criticized on the grounds that ethical statements appear to represent synthetic *a priori* propositions, the legitimacy of which most philosophers deny.

It is precisely at this point that the moral/ethical distinction becomes significant. The content of a statement such as "sexual congress with children is immoral" is not known to be true intuitively. It does not possess the characteristic of incontrovertibility. However the *conviction* that such behavior is morally wrong is an evaluation that is performed as a cognitive process and an affective experience, just as one senses that if $A = B$, and $B = C$, then $A = C$. "Abuse" of a child is considered immoral simply because "injury" to human life is commonly believed to be a transgression on the moral evaluative scale. Although the "injury" in any specific instance is subject to interpretation, the precious nature of life is a universally accepted concept by civilized people.

Provincialism

Among the many who cannot accept the notion of an intuitively known moral sense are those known as *provincialists*. Their principal disagreement has to do with claims regarding the potential uniqueness of life. They do not propose an alternative explanation for the source of moral feelings but rather deny the validity of any analysis that presumes that living beings are characterized by any such emergent property. Provincialists contend that biology is a province of physical science and can advance "only by applying the methods of physical science, and nowadays especially, the methods of physical and organic chemistry."[34] Physics is, of course, considered the epitome of scientific analysis. "Indeed for most of us," said Rosen, "it simply *is* science."[35]

Kauffman referring to the discovery of the structure of DNA, and the expression of genes as proteins and as regulators, etc. said: "It is here that the program of reductionism [i.e., provincialism] in biology has been most profoundly successful."[36] There is certainly no reason, he argues, for the positing of a moral sense that is peculiar to life. Wolff agreed, proposing that "morality is divorced from economic, biological, psychological, social and political features of man and society. It is as relative as the culture that contains it and as arbitrary, irrational, and idiosyncratic."[37] While this appears a damning denial of the existence of a moral sense, it is clearly more accurately related to ethics, which represents the learned manifestation of the moral sense. Reductionist interpretations are, unfortunately, a class of interpretations that are both inaccurate and misleading.

Those who oppose reductionist interpretations (*autonomists*) argue that the aims as well as the research methods of biology "are so different from those of the other sciences that biological theory and practices must remain permanently insulated from the distinctive methods and theories of physical science."[38] This diametrically opposed interpretation has been offered by Rosen and others who believe that physics may be too narrow a science to embrace biology. They believe that it may be more appropriate to reduce physics to biology, since the principles of biology are more general. Levins and Lewontin, who propose a dialectical view of nature, said that "despite the extraordinary success of mechanistic reductionist molecular biology, there has been a growing discontent in the last twenty years with simple Cartesian reductionism as the universal way to truth."[39] Boden, for example, proposed that:

The reductionist temperament shows itself in such assumptions (for instance) as that Neo-Darwinism can answer all evolutionary questions, molecular biology all questions about individual morphogenesis, stimulus response psychology all questions about behavior, and neurophysiology all questions about the mind.[40]

One's view of the nature of ethics must certainly be influenced by the acceptance of one or the other of these positions. A provincialist interpretation would deny that moral behavior exists. Obligation, duty, responsibility, etc., would have no meaning. The autonomist view is that culture and attendant learning ultimately produce individuals that are, at least to some extent, free of the strictures of chemico-physical law. The latter interpretation is sympathetic to the acceptance of purpose as a characteristic of the living, which by its nature begets altruism.

Waddington agreed, proposing that "purposes are...not only something which it is proper for scientists to discuss, but they actually demand discussion."[41] Grene took a similar view. While recognizing that according to the scientific community all discourse "must be wholly non- and anti-teleological,"[42] she pointed out that:

> this ideal...is, to say the least, impracticable, because in order to decide what facts to study one must understand a great deal about the orderly development of the organism.... Teleological discourse has not only a regulative but at least a *descriptive* function within biological research.[43]

Many philosophers of science take the same position. Rosenberg observed that "the immunologists willingness to employ terms like 'recognition' to describe cell-to-cell interactions, and the molecular biologist's appeal to 'codes,' 'information,' and 'error,'...to describe DNA transcriptions are remarkable."[44] And, he added, "nothing is more striking in biology than the apparently goal directed phenomena of embryology and development."[45]

The Good and the Natural

Along with many other arguments regarding the nature of morality is the question of whether *natural* behavior is necessarily *good* behavior. Is what one does on the basis of instinctive urges always an example of what one *should* do? From a sociobiological perspective, anything done on the basis of what abets one's lineage qualifies as "good" to the extent that it is fitness enhancing. While a behavior may appear "bad" to an observer, it may, in fact, be motivated by sentiments created through genetic influence. It is perhaps not merely a desired, but, in fact, an essential step in a survival sequence.

Psychologists such as Campbell are appalled at such an interpretation. He stated his unequivocal rejection of the view that "what is biologically natural is normatively good."[46] He upbraided psychoanalysts for what he saw as their consistent attempt to justify narcissism and to condone such behavior as insurrection in depressed areas by "justify[ing] such rioting, giving it a positive moral value."[47] He concluded that to the extent that human survival is accepted as an ultimate value, it is possible to "designat[e] certain *innate* (italics added) human behavioral tendencies as sinful or non-optimal."[48]

Campbell based his argument on the fact that environmental changes have made previously adaptive behavior (e.g., eating copious amounts of sugar, fats, and spices) maladaptive. Insofar as he was referring to the fact that changes in the environment are responsible for causing a behavior to become maladaptive, there is little room for disagreement. However, when he referred to *innate* tendencies as sinful, he was suggesting that the sin lies in the tendency. He failed to appreciate the distinction between *desires* which are innate, and *behaviors*, most of which are learned. Immorality lies in the performance of an iniquitous act, not in the desire that prompts it, lest people be judged evil for the experiencing of carnal desire.

Even with that reservation, it is difficult to understand the labeling of an act immoral where it does not bear on reproductive potential. Solomon reported on the conclusion drawn by a group of professionals regarding the relationship between natural behavior and morality among lower animals, offering the following observation:

> It would in no way be presumed [by sociobiologists] that the survival of [a] species or its basic genotypes was a "good" in any but the trivial sense of "having made the survival of the species possible." But when the species in question is homo sapiens, the survival of the species or its basic genotypes suddenly appeared as just such a "good," in fact, an ultimate good, and the trait in question accordingly appeared as a virtue.[43]

Solomon suggested that such an interpretation represents an "impurity," and he wondered whether "sociobiological study of human organization could be purged of such a presupposition."[50] It is difficult to see how he could come to such a remarkable conclusion. Solomon is a philosopher, and, in particular, an ethicist. It is clear that he rejects sociobiological interpretations of ethics on a number of grounds. However, there emerges a sense that he (Solomon) believes that *he* knows what morality is—or at least that he knows what it isn't. However, no alternative is offered. There is only his explicit conviction that the nature of the "good" certainly isn't what sociobiologists say it is.

Solomon's further analysis of morality as including nonaltruistic behavior is similarly distressing. He stated that such rules as those enumerated in the ten commandments call only for following the law; that rituals of "correctness," and Buddhist notions of earning "merit," require no moral or altruistic—only conforming—behavior. But what is the basis for such codification? The fact is that every such case of constraint on the expression of self-gratifying urges can be shown to represent the acceptance of the morality residing in respect for others (as well as for the sanctity of one's own body); *a respect that can be expected to call for altruistic behavior should the occasion arise.* This is not to contend that all examples of what appear to be ethical behaviors mentioned are morally correct, but only that they can be evaluated on the moral scale.

The Emergence of Moral Behavior

In order to deal with the derivation of normative ethical theories the question of the basis on which philosophers come to construct their models and the role that morality plays must be addressed. Halevy suggested that "the task of the moralist is to analyze this [moral] feeling, to define what moral feeling actually is."[51] He stated that a first step should be to determine what the function of a moral sense might serve; what such an affective state may signify; from whence it may have arisen. *That is the issue to be addressed in this text.* Mandelbaum et al. suggested that "all philosophers have agreed that a system of ethics must give some account of the pretheoretical moral beliefs of mankind."[52] They contended that ethical theory building should be seen as a process by which general laws are developed on the basis of what is observed to be common practice.

And what is it that is observed? It is quite impossible to attend to the daily behavior of millions of people of all ages, cultures, creeds, races, nationalities, etc., without being aware that moral feelings abound. Lecky summarized the view of historians and sociologists, saying that "men feel that a certain course of life [i.e., ethical behavior] is the natural end of their being, and they feel bound, even at the expense of happiness, to pursue it."[53] But why do individuals *ever* risk their own well being, or sacrifice their personal treasure in the interest of aiding others? The response that many biologists and other professionals consider naive, though it is fostered by many religious institutions, is that man is invested with a sense of obligation to the welfare of others. "We are our brother's keepers" they contend, and they contend that all men are brothers. The scientific community, however, finds such explanations nonsensical.

Some argue that altruism came to be practiced as a defense against threat, with most theorists reserving moral behavior to (socialized) human societies. Alland, for example, proposed that "morality can only have a value for survival after man *qua man* has emerged and the human type of sociocultural system has been established. [It] exists in man as a byproduct of both his genetic structure and his social existence."[54] And Warnock proposed that "the 'general object' of morality...is to contribute to betterment—or non-deterioration—of the human predicament.... It is the proper business of morality...to make us more rational in the judicious pursuit of our interests and ends...to expand our sympathies, or better, to reduce the liability to damage inherent in our natural tendency to be narrowly restricted."[55]

Ardrey agreed that the need for a defense against threat forced people to become altruistic—to establish peaceful relationships. In this, he contended that human behavior mirrors that of simpler animals. "The amity...which an animal expresses for others of its kind will be equal to the sum of [conspecific] forces arrayed against it."[56] Caporael et al. support this hypothesis, saying that "sociality was a prime factor shaping the evolution of *Homo sapiens*. The cognitive and affective mechanisms underlying such choices evolved under selection pressure on small groups...for controlling the behavior of other group members."[57]

The biological aspect.

For those taking the position that altruism is a peculiarly human phenomenon, considerable difference of opinion exists as to the level of evolution at which it is most appropriate to assume that human characteristics appeared. Anthropologists have been engaged in a running battle over the interpretation of data collected from fossils. Brain size, dental characteristics, jaw shape, total body size, dimorphic male/female features, etc., have all been employed to make a case for the preeminence of conflicting viewpoints. Leakey contends that the emergence of true humans was precipitated by the advent of the genus *homo*, an offshoot of the genus

Australopithecus afarensis, which many believe "was the ancestral stock from which all hominid species evolved."[58] That step is believed to have occurred approximately 7.5 million years ago. However, he contends that the first such hominids, or protohumans "can best be described as bipedal apes."[59] He does not consider them to have been truly human.

The reason for this interpretation is that the earliest such individuals were more apelike in many of their characteristics (small brain, large cheek teeth, etc.) than human, although the difference between apes and humans "in the basic genetic blueprint, DNA, is less than 2 percent."[60] Stebbins had pointed out some time earlier that "with respect to any single characteristic that one might...analyze in depth, humans differ only quantitatively from apes."[61] He quoted Baumgartel who said: "Gorillas are like us in so many ways.... They love, protect, care for, and discipline their children.... Love is an essential emotion in the lives of these animals."[62] It is difficult to imagine an explication of love as it is defined by Baumgartel that does not include reference to altruistic interactions.

The appearance of a genus that seemed to meet the conditions essential to being considered human occurred about two million years ago. Leakey suggests that, *"Homo erectus* seems...to be at the threshold of something important in our history."[63] His contention is that "more or less everything that preceded *homo erectus* was distinctly apelike in important respects: in some of the anatomy, life history, and behavior. And everything that followed *erectus* was distinctly humanlike."[64] Thus for Leakey, a clear distinction can be drawn at some point in the history of animal life.

The method by which this "humanness" emanated is a widely disputed issue. Until recently it was assumed that as populations of *homo erectus* individuals spread across Europe (the "old world") they passed through a number of intermediate evolutionary stages. Finally, through gene flow and interspecies contact, *homo sapiens sapiens* emerged. Because under that interpretation it is assumed that the evolutionary process was carried out in many places in a similar fashion, it is known as the multiregional hypothesis. An opposing view, popularized in the 80's, is that the *homo sapiens sapiens* branch of the hominid family arose as a single event—a speciation—at one place, from which it spread across Europe by dominating and ultimately replacing more primitive types through superior selection attributes. This is known alternatively as the "Noah's Ark," "Garden of Eden," and "Mitochondrial Eve" model.

The evidence for this "single emergence" theory is based on a study of mitochondrial RNA, which is inherited only through the female line. The property of mitochondrial DNA that makes it useful to researchers is that it mutates very rapidly. Thus, a study of extant humans can reveal how much genetic alteration has occurred and, how far back in time the human species probably came into being. Researchers have determined that "the overall degree of genetic variation in the mitochondrial DNA of modern populations is modest, which implies the relatively recent origin of modern humans."[65] And since African populations have the "deepest genetic roots," it is assumed that Africa is the source of all human populations, having arisen approximately 120,000 years ago, which supports the single event theory.

Unfortunately, in spite of such evidence, paleontologists and anthropologists are not fully convinced. Fox argues that "a large amount of fossil evidence has come to light documenting the gradual transition that, even in the absence of direct evidence, Darwin realized *must* have happened."[66] And the record of evolution shows no sharp break between man and animal that can be pinpointed at a certain brain size or anything else."[67] This would suggest that a moral sense was unlikely to have emerged suddenly. Leakey agrees, concluding that as uncertain as we are about the first use of tools, and the appearance of art forms, etc., "least certain of all is an understanding of the precise evolutionary change on which modern human[ity] is founded."[68]

The significance of this issue lies in the fact that the "essence of humanity" which perhaps involves the evolution of a moral sense is shrouded in mystery as to its beginnings. It would be extremely difficult to determine whether animal "moral analogous" behavior was ever more than that. From the point of view of moral development the critical questions are: What are these characteristics that may bear on the emergence of altruistic behavior? Why should humans act any differently toward fellow members of their species than did members of earlier species? Why should not the principles of reciprocal altruism, (risking in the hope of later benefit) and perhaps an extension of the practice of kin selection, (sacrificing for one's close relatives) be followed?[69]

Leakey's explanations are not particularly convincing. His position is that with an increase in brain size and mental capacity emerged "the real beginning of the burgeoning of compassion, morality, and conscious awareness that today we cherish as marks of humanity."[70] He argues that more cooperation and friendly interaction occurs between members of *homo* troops than among more primitive species because:

> "the males are related to one another; brothers, half brothers, and so on.... Bonds between kin are strong...but strong friendships and al-

liances form too, sometimes as male-female cohorts, but more often as 'political' alliances.... When homo males reach maturity, they don't transfer to another troop... but stay in their natal troop, brothers and cousins living together, cooperating with one another."[71]

He concludes, in agreement with Hamilton, that kin selection has acted to select families in which altruistic behavior occurs.

More significant is the evidence that in *homo sapiens*, birth occurs some time earlier in the "life history" of the individual than it does in earlier life forms. "The size of the human pelvic opening...has increased during human evolution...but there are engineering constraints that limit the size of the birth canal, constraints imposed by the demands of bipedal locomotion."[72] What this has meant to the evolutionary process is that the human infant, being "born too soon," is more docile to learning than are more primitive forms of life, and social learning, a precursor of altruistic behavior, has become a most vital developmental element. Markl made the point that through an analysis of the ontogenetic process in children may be found "a model for a process which must have occurred in evolution in the phylogenetic transition from animal to man."[73] The study of infant behavior should provide valuable information regarding the development of the moral sense, both phylo, and ontogenetically.

Are humans basically "good" or "evil"?

One of the most confusing aspects of the moral issue has to do with basic assumptions regarding the fundamental nature of humans. Are people inherently good, with evil representing the impact of negative social experiences, or are they essentially depraved, requiring the strictures of ordered community life to hold them to a narrow, respectable, course? Obviously, such questions involve reference to a moral sense. However, the role that such an affect is believed to play varies widely. Philosophers, psychologists, and theologians have taken totally disparate views, with each including representatives at both extremes. Aquinas contended that humans are essentially good (though the Christian Bible paints them as inherently bad). Metchnikoff said that "while rational philosophers in all ages have...held human nature to be good...many religious doctrines have displayed a very different view. Human nature [is] regarded as being composed of two hostile elements, a body and a soul."[74] He went on to describe the many self-mortifying practices that are carried out in religious ceremonies as evidence of an awareness and abhorrence of inherently evil tendencies.

Rogerian and most other phenomenological and existential psychologists appeal to a positive human nature. Fromm stressed that the behaviors of emotionally healthy people are "rooted in the bonds of brotherliness and solidarity."[75] Psychoanalytic psychologists, by contrast, have usually described individuals as inherently cruel and selfish. Freud argued that "man's natural aggressive instinct...opposes [the] programme of civilization."[76] Menninger referred to "the destructive instinct that slumbers within the heart of even the tiny child,"[77] and Anna Freud regarded infantile sexuality as "of a purely perverse nature."[78] Taking an eclectic approach, Mill said:

> The difference between a bad and a good man is not that the latter acts in opposition to his strongest desires; it is that his desire to do right, and his aversion to doing wrong are strong enough to overcome...any other desire or aversion which may conflict with them.[79]

Many interpretations are, knowingly or not, consistent with the thesis of sociobiologists; the notion that humans as well as animals are necessarily selfish and that the labeling of any behavior as altruistic represents no more than a technique for masking true intentions. But how shall the accuracy of such contentions be determined? Metchnikoff proposed that morality should not be based on existing, but on ideal human nature, and this task, he said, "can be undertaken only by science."[80] Cattell agreed, expressing the view of scientists from a variety of disciplines:

> Morality has to do with goals, and if a desirable ultimate goal could be found for mankind it should be possible to hand to social science—the more mature science of tomorrow, if not the feeble infant of today—the task of finding what behaviors in individual men best brings us nearer to that goal.[81]

Although not all scholars would agree on whether such goals are self or other-oriented or that the social sciences are the appropriate branch, sociobiological study is believed by many to represent a significant bough of the scientific tree.

Altruism and Social Convention

Many of those who deal with the concept of ethics separate altruism from other behaviors that govern social interaction (e.g., customs, etiquette, laws) arguing that the latter represent no more than sets of rules established in the interest of customs peculiar to various communities and sanctified by time. Taylor asked: "How is the moral code of a society to be distinguished from its legal code or its code of etiquette?"[82] (Obviously assuming that some difference between them can be discerned.) A legal code, however, must surely be based on moral principles.

Codes of etiquette seem less clearly related. Laird contended that "politeness is obviously a social affair; yet is usually accounted a very minor part of morals, although socially of indisputable importance...[but] it is a mistake to say that any question is moralized precisely in proportion as it is socialized."[83] He contended that the fact that a particular behavior is a common social practice does not guarantee its moral quality. Turiel presented a more extreme view, stating that "in themselves social conventional acts are arbitrary, in that they do not have an intrinsically prescriptive basis: Alternative courses of action can serve similar functions."[84]

He went further. He took the position that "The failure to distinguish between these two questions, between morality and social convention, has been a major obstacle in the social-scientific study of morality."[85] Kowalski agreed, claiming that "moral rules are to be distinguished from at least two other sorts of rules: laws...and etiquette, or social conventions."[86] He added the observation that Turiel had shown that "three to four year old children can distinguish moral rules and social conventions."[87]

Each of these authors is right, in their concern regarding the problem of the relationship between morality and social convention. *However, they have the case backward.* The obstacle to an understanding of morality and its attendant ethical behavior, is based, in part, on the inability to recognize their *common* basis. The issue that is relevant to this text is, in fact, the specific relationship between social practices and altruism. The question to be dealt with is why ethical actions become legitimized. How do societies determine which actions are appropriate and which are not?

The intrinsic factor is the moral sense, while the regulations involved are those that characterize both overtly altruistic behavior, and ethical practices which are implicitly altruistic. No case whatever is made for the universalizability of ethics (which is at the heart of their concern). Justice, for example may be expressed at different times and in different places in a wide variety of ways. However, the moral sense which animates both altruistic behavior and the various socially mandated actions that reflect it is an innate characteristic of the human psyche.

The Dilemma

It is a constant theme of approaches to human adjustment that attempts should be made to assist people to function in ways that are consistent with the dictates of their genetic endowment; to accept, for example, the positive aspects of sexual behavior, the normalcy of aggressive activity, the function of grief, anger and other emotional responses to human experience. If effective guidance is to be provided—especially for developing children—it is imperative that programs be based on the best available evidence, and the model provided by sociobiologists will be abjured in its radical interpretation here. The thesis to be developed is that all behavior is based on decisions that mandate the sacrifice of the expression of some desires in the interest of gratifying others, and that on some occasions the welfare of others will take precedence over that of the behaving individual.

Of course, most attention is paid to the satisfaction of more immediate desires—regardless of the risk to one's ultimate well being. The excessive use of sugar, fats, and other comestibles are apt to be ruinous to health. The existence of a surfeit of appetite satisfying foods (fast food chains base their success on the adroit mixture of salt, grease, and sugar) is a product of the explosive growth of the ability to refine, extract, combine, and distill substances in order to appeal to one's smell, taste, and other senses. Other products of civilization are the graphic display of sexually erotic and often pathological forms of expression, and the tantalizing of millions of viewers with violently aggressive behavior and perverse forms of sensualistic stimulation.

It is, in fact, considered perverse to become wholly other-directed; to lose one's sense of self in the interest of aiding others. The concept of "codependency" or the practice of extending so much caring and support for individuals with drug addictions and other problems that the "helper" suffers psychologically has become an important factor in therapeutic programs. Religious zealots that sacrifice their health—and sometimes their lives—in bringing the word of a spiritual leader to the populace are often labeled schizophrenic. There is, however, a middle ground.

Psychologists, philosophers, social scientists and other professionals must deal with the question of whether such desires as those for compassionate experiences, for contributing to the welfare of a community, for sharing and otherwise sacrificial behavior, are genetically based. Do they represent a unique class of desire, or are they merely a series of techniques for deriving profit to the behaving individual, and, through the procreative process, to a specific gene line? If the latter is the case, it would seem far more efficient to teach children the principle of "enlightened self-interest" which is not only a common practice in many families and in many societies, but is consistent with the principle of reciprocal altruism—of behaving in a manner that will bring some ultimate profit to the behaver.

However, such a program would, according to the philosophy to be developed here, represent an ignominious retreat from social responsibility. Even the most rabid sociobiologists propose the teaching of altruistic behavior.

Summary

From time immemorial philosophers have attempted to understand the nature of morality and its influence on behavior. Until the very recent past, and before the many major discoveries in the fields of genetics and molecular biology, attention was focused on the field of normative ethics. The critical questions involved which behaviors are most morally appropriate; how people should live their lives; what is the nature of the "good" and the "evil." The major struggle was between those who contended that humans are no more than a special case of chemico-physical nature, and those who argued that only humans have an "intuition" that accounts for moral convictions.

Today many of these questions are considered anachronistic. Scientists of many disciplines have adopted a view of morality, and particularly of altruistic behavior, that has radically altered the nature of the topic. Sociobiologists focus on the distinction between assisting one's relatives and sacrificing for strangers. They have concluded that all behavior is selfish, in that it serves only gene line perpetuation. In spite of the claims of these professionals, there is considerable evidence to suggest that humans possess a moral sense that guides much of their behavior. The acceptance of one or the other of these diametrically opposed interpretations of altruistic behaviors should have implications for many social and political decisions. The challenge of this and similar texts is to make an effective case for the influence of a moral sense, while taking into account the powerful evidence provided by sociobiologists for the primacy of the gene.

The Issue

Chapter 1 Notes

1. Socrates referenced in Russell (1945), p. 573
2. Spinoza (1903), p. 172
3. Flew (1979), p. 113
4. Runes (1962)
5. Rule utilitarianism is that branch of utilitarian philosophy that stresses one's obligation to perform certain behaviors regardless of the consequences of that particular action, on the assumption that if all people acted in such a manner the *net* outcome would be positive.
6. Flew (1979), p. 112
7. Runes (1962), p. 202
8. Kohlberg (1987), p. 292
9. Albert & Denise (1988), pp. 5-6
10. Waddington (1971), p. 42
11. One might say that ethics relates to morality as an apple pie relates to the *sense* of taste—perhaps more accurately to the sense of sweetness.
12. Wonderly (1991), p. 241
13. Moore (1965), p. 321. He said that Plato risked defining good by saying that "a creature that possesses [good] permanently, completely and absolutely, has never any need of anything else; its satisfaction is perfect," (*Ibid.*).
14. Hare (1981), p. 55
15. Warnock (1971), pp. 1-2
16. Prichard (1975), p. 392
17. Sharp (1928), p. 75
18. Baier (1964), p. 291
19. Consider another of Baier's contentions. "The very *raison d'etre* morality is to yield reasons which overrule the reasons of self-interest in those cases when everyone's following self-interest would be harmful to everyone" (1958, p. 308). No such reasons will be discovered, since they would all be reducible to self-interest.
20. Markl (1980), p. 220
21. Runes (1962), p. 10
22. *Ibid.*
23. Brandon & Burian (1984), p. xiii
24. Dawkins (1982), p. 57
25. Williams pointed out that altruism is sometimes misunderstood to represent a motive. "It is meant, in fact, to be the name of a function; that is, there is supposed to be a feature of the institutions of morality which has

the effect that other people's interests get observed...and this effect is no accident," (1980, p. 276).
26. Many sociobiologists have proposed such an interpretation. Rachels, for example, suggested that "we may also use the word in a somewhat stronger sense, as involving the *willingness to forego some good for oneself* in order to help others," (1991 p. 149). (Note that his use of the term "willing" refers to an affective state related to the payment of a price in the interest of gaining some desired outcome.)
27. Reid (1965), p. 291
28. Huxley (1900), p. 223
29. Plato referenced in Copleston (1962), p. 114
30. Ruse (1988), p. 72
31. *Ibid.* p. 73
32. Hume (1938), p. 109
33. Laird (1970), p. 63
34. Rosenberg (1985), p. 16
35. Rosen (1991), p. 117
36. Kaufman (1993), p. 8
37. Wolff (1974), p. 221
38. Rosenberg (1985), p. 16
39. Levins & Lewontin (1985), p. vi. Roszak was similarly distressed, referring to reductionism as "that particular sensibility which degrades what it studies by depriving its subject of charm, autonomy, dignity, mystery," (1972, p. 263). Koestler referred to it as a "pillar of unwisdom" and Maslow called it a "cognitive pathology." Many others have voiced the same opinion.
40. Boden (1984), p. 318. Rosenberg pointed up the vehemence of the disagreement between opposing schools, saying that "autonomists often rail at reductionists as though it were a term with the moral connotations of 'racism' or 'fascism,' (1985, p. 19). Autonomists contend that it may be that living creatures represent not an insignificant perturbation in the flow of inanimate existence, (certainly one interpretation in terms of their insubstantial existential proportion) but a class of entities whose nature cannot adequately be revealed through physical and chemical analysis. Rosen asked: "Why could it not be that the "universals" of physics are only so on a small and special (if inordinately prominent) class of material systems, a class to which organisms are too *general* to belong. What if physics is the particular, and biology the general, instead of the other way around? (1991, p. 13).
41. Waddington (1971), p. 26

42. Grene (1974), p. 175
43. *Ibid.*
44. Rosenberg (1985), p. 133. As to what end is served, what purpose carried out, he added: "A function may be adaptive for the organism that manifests it...or the organisms kin group...or, more controversially, its species or ecosystem," (*Ibid.* p. 48). Thus, he believed that true altruistic behavior is a possibility.
45. *Ibid.* p. 37
46. Campbell (1980), p. 70
47. *Ibid.* p. 71
48. *Ibid.*
49. Solomon (1980), p. 257
50. *Ibid.* p. 258
51. Halevy (1966), p. 12
52. Mandelbaum et al. (1967), p. 525
53. Lecky (1955), p. 181
54. Alland (1971), p. 58
55. Warnock (1971), p. 26
56. Ardrey (1966), p. 219. Others, though not specifying the uniqueness of humans as moral beings, agree that many altruistic activities are practiced as a response to danger—especially as represented by the behavior of other humans. Warnock said that "moral judgment is a practical matter; it is supposed to make, and often does make, a practical difference," (*Ibid.* p. 26). Morris contended that the threat of the external world, and its many different—sometimes hostile—people resulted in "a basic shift toward mutual aid, towards sharing and combining resources," (1969, p. 240). Each individual created a relationship with some small "tribe sized" group within which "he could satisfy his basic urges towards mutual aid and sharing," (*Ibid.*, p. 25). Gough provided an example in his description of behavior in early Greek communities, saying that "everyone...was in danger of being injured by his neighbors, and therefore, men made a contract or bargain...with one another, by which each man undertook to refrain from injuring his neighbor, provided that his neighbor would refrain from injuring him," (1957, p. 10).
57. Caporael et al. (1989), p. 683
58. Leakey (1992), p. 115
59. *Ibid.* p. 92
60. *Ibid.* p. 83
61. Stebbins (1982), p. 363
62. *Ibid.* pp. 362-363

63. Leakey (1992), p. 55
64. *Ibid.* p. 64
65. *Ibid.* p. 221
66. Fox, R. L. (1989), p. 13
67. *Ibid.* p. 29
68. Leakey (1992), p. 236
69. The terms "reciprocal altruism" and "kin selection" were defined specifically in the introduction, p. xvii.
70. Leakey (1992), p. 157
71. *Ibid.* p. 140
72. *Ibid.* p. 160
73. Markl (1980), p. 214
74. Metchnikoff (1977), p. 9
75. Fromm (1973), p. 362
76. Freud, S. (1961), p. 122
77. Menninger (1938), p. 24
78. Freud, A. (1949), p. 39
79. Mill (1957), p. 50
80. Metchnikoff (1977), p. 289
81. Cattell (1972), p. 76
82. Taylor (1964), p. xiii
83. Laird (1970), p. 80
84. Turiel (1980), p. 110
85. *Ibid.*
86. Kowalski (1980), p. 233
87. *Ibid.*

Chapter Two

Conflicting Views

> It is of no little moment for the future whether people are necessarily and consistently selfish...or whether there is a significant place for altruism in the scheme of human behavior.
>
> *Herbert Simon*

Beliefs regarding the nature of ethics are spread across a wide spectrum, ranging from the extreme view of creationists who insist that evolutionist views in general, and sociobiological claims in particular, cannot account for the moral behavior of humans, to the majority of life scientists, who reduce all deliberate action to selfish motives. Between these extremes lie those biologists who, though they accept many of the principles of genetic determination, deny that people are fated to submit to the demands of the genome in every aspect of life. Such individuals apparently lean heavily on the intuitive sense mentioned earlier.

Many biologists castigate notions of inherent senses. Wilson says that "the Achilles heel of the intuitionist position is that it relies on the emotive judgment of the brain as though that organ must be treated as a black box."[1] Sociobiologists believe that they can unseal that box. Wilson, like others in fact, offers an answer to the question of how people are encouraged to engage in cooperative activity. It is accomplished, he says, "simply by incorporating morality within the innate dispositions."[2] But this concerns such philosophers of biology as Ruse who argues that "since sociobiologists believe that all actions are ultimately selfish, the embracing of 'innate dispositions' means that morality reduces to self-interest. And this, of course, is not true morality."[3]

Sociobiology:
From Genetics to Genaddicts[4]

With the advent of molecular biology a revolution in evolutionist biological thought took place. Throughout the first decade following the discovery of the double helical structure of DNA, the influence of the breakthrough was confined essentially to professional journals, although some of the implications began to spread to neighboring fields, particularly physics, as reductionists saw the implications, and to neo-Darwinists, who recognized the support that genetic discoveries could provide. Crick contended that "the ultimate aim of the modern movement in biology is...to explain all biology [life] in terms of physics and chemistry."[5]

In the 70's a number of life scientists caused a minor eruption as they brought the topic into the mainstream of public attention. Hamilton, Trivers, Dawkins, Wilson, and others wrote, interpretations including the social implications, of molecular biological findings. Some dealt with the behavior of insects, some with that of mammals, and some with genetic influence on human ethics. At the extreme they called for an acceptance of the notion that every ostensibly altruistic act must ultimately profit the behaver or the behaver's kin, or at least convince that individual that reciprocity is apt to follow. "As a general rule, a modern biologist seeing one animal doing something to benefit another assumes either that it is manipulated by the other individual or that it is being subtly selfish."[6] No matter the ostensible genuineness of an unselfish act, it is proposed that in the final analysis all behaviors must be understood as providing for the continuation of the individual's genes in succeeding generations.

This reductionist aspect of biology (sociobiology or evolutionary biology) is a field of study representing an outgrowth of the ethological movement of the 1950's in which the behavioral aspects of evolutionist theory became the object of scientific analysis. A majority of biologists have applauded sociobiological concepts. The acceptance of some of the major principles appear to represent an opportunity to gain the kind of respectability ordinarily reserved to the "hard" sciences of physics and chemistry. Dyson, for example, referred to Dawkins' analogy between genetic and cultural evolution as having been "brilliantly explored."[7] Molecular biologists, said Kaye, are similarly assumed to be basing their conclusions on the "unambiguous social meaning of their scientific work."[8] The conclusion being drawn is that "from being the ugly duckling of the evolutionary family, the study of social behavior [has become] a very proud and magnificent young swan indeed."[9]

As a philosophical matter, sociobiologists are provincialists or ontological reductionists, being convinced that chemical and physical laws account for all biological phenomena. They "make a firm commitment to the ultimate reality *only* (italics added) of material particles."[10] Furthermore, only individuals bearing the "fittest" genes can be expected to reproduce. Thus the welfare of the group is never of greater significance than that of its individual phenotypes. Ruse described the position:

> The organism with an adaptation of benefit to the group, but which adaptation in some sense counts against itself, is going to be at a selective disadvantage to an organism with an adaptation of benefit to itself, and not at all...to the group. The second organism will out-reproduce the first organism.... Consequently, in the next generation only the second organism will be represented, and however much good the first organism's group-benefiting characteristics may have been, they are gone and lost forever.[11]

Wilson suggested that when moral development and genetic variance problems are clarified, "the two approaches can be expected to merge to form a recognizable exercise in behavioral genetics."[12] He went further, proposing that "the time has come for ethics to be removed temporarily from the hands of the philosophers and biologized."[13] Ghiselin took an extreme position. In a cynical denial of truly other-directed behavior he said, "Scratch an 'altruist' and watch a 'hypocrite' bleed."[14]

Trivers coined the term "reciprocal altruism" introduced above, suggesting that generous actions are performed in the interest of acquiring "credits," to be cashed in when something is needed by the benevolent initiator. The case for reciprocal altruism as it was developed by Hamilton (1964) and Trivers (1971) was based on an analysis of instances of extreme sacrifice.[15] Most extreme, however, was Dawkins concept of the "selfish gene."[16] In this (1976) publication, the interest of genes was seen as supreme. Only in the case of closely related individuals (kin selection) could altruistic behavior make sense. Solomon agreed with Dawkins' argument, pointing out that "it is the survival of the genotype that is the key to sociobiological theory."[17]

Although he is recognized as a leading spokesman for the synthetic evolutionary theory Mayr has not accepted the genetic thesis unconditionally, saying,

> "I do not believe that definite genes controlling character traits of high ethical value have ever been demonstrated. Rather what is inherited are tendencies, and capacities.... The evidence indicates that the genetic component in human ethics is...of minor importance."[18] [His thesis was that individuals have a capacity for learning ethical behavior.] People "can adopt a second set of ethical norms supplementing, and in part replacing the biologically inherited norm based on inclusive fitness."[19]

This capacity, he apparently assumed to be extragenetic, though the source is not indicated.

Support for the sociobiologist's position can be found in the views of many philosophers. Hobbes, for example, held that "of the voluntary acts of every man, the object is some good to himself."[20] But, he said, there is no security in such behavior. "Thus, the second law of nature is that he shall think it necessary to lay down this right to all things; and be contented with so much liberty against other men, as he would allow other men against himself."[21]

The thrust of these power based philosophies is that morality is either a form of weakness (e.g., Nietzsche's slave morality) or a form of behavior that represents a compromise designed to maximize benefit to the individual. A wide range of scientists is coming to support the view that sociobiologists suggest. "Today," says Kaye, "natural philosophers find only confirmation of [God's] absence."[22] They are overwhelmed by the knowledge displayed by genes that seem to have the capacity to determine every aspect of behavior.

Sociobiology and the self

Sociobiologists contend that the goal of all behavior is ultimately the enhancement of the *self*, which includes both the physical individual and that individual's lineage. The normal state of man is to be completely selfish; to provide the opportunity for mindless genes to propogate. "So long as DNA is passed on, it does not matter who or what gets hurt in the process. Genes don't care about suffering, because they don't care about anything."[23] Translated into human behavior, interpretations are often obnoxious. A well known literary critic, exemplified the process bragging that:

> To me, pleasure and my own personal happiness... are all I deem worth a hoot. It would make me out a much finer and nobler person. I duly appreciate, to say that the happiness and welfare of all mankind were close to my heart, that nothing gave me more soulful happiness than to make others happy and that I would gladly sacrifice every cent I have in the world, together with maybe a leg, to bring a little joy to the impoverished and impaired sufferers of the late floods in India, but I have difficulty in being a hypocrite.

He believed that all men share his view. Bernard de Mandeville, a psychological hedonist, not only agreed with that position but contended that society is better off because people look out first for themselves. De Mandeville, in fact, argued that there is no merit in saving a child ready to drop into a fire. He suggested that such action is neither "good" nor "bad", and what benefit the infant would receive from being rescued is a result of the fact that our failure to take action would have caused us pain, which self preservation compels us to prevent.

This view of the self as being a purely self serving entity, has been championed in every century, and by many philosophers. Hume disagreed, however, stating that the self is no more than a product of the "mental" states. "The ultimate realities are the mental states, and the selves are only secondary, since they are nothing but aggregates of the states.... We must no longer say that the self perceives, thinks, or loves, or that it has a perception of thought or an emotion."[24] Mead proposed that the self is a social phenomenon.

> The self has a character which is different from that of the physiological organism proper.... It is not there, at birth, but arises in the process of social experience and activity. That is, it develops in the given individual as a result of his relations to that process as a whole.[25]

Such views suggest that the self is an independent entity. It is, however, only the focal point along a continuum of awareness; the biological individual being at one extreme and the totality of a gene pool across time and space at the other. It is a meaningful concept only at the level of life at which it is possible for the individual to act counter to the survival and growth potential of the parent gene pool. In the simplest organisms drive information is spread evenly across the range of possible responses. Urges toward eating, sexual activity, and stimulation all bear on the individual as impulses to action and arouse similarly powerful emotions when relevant situations arise. When organisms capable of affective experience became involved the nature of the self was altered.

The problem under investigation here is the insistence with which many biologists hold to the view that selfish behavior is an inevitable result of the fact that only genes or gene groups are units of selection; the denial of the possibility of group selection.

Culturalism:
Biotheology or Theobiology?

If sociobiologists are greatly impressed by the evidence coming from the research of molecular biologists and organic chemists, their detractors in the scientific community are equally adamant in their denial of such claims. Although most biologists today are supportive of genic selectionism, a significant number of scientists find such explanations anathematic. They believe them to be much too simplistic. Waddington, for example, took the position that "a causal explanation of evolution cast purely or even primarily in terms of genetics will be incomplete [in that] a more complete theory of evolution entails complexity and subtlety that cannot be encompassed by a genetic reductionism."[26] Unfortunately he said, the most extreme advocates of the sociobiologist viewpoint "are adamant that natural selection must be brought to bear on all human social thought and behavior."[27]

Wolff contended that sociobiologist insistence that genetic control of behavior is inflexible "should be challenged on moral as well as scientific grounds."[28] He expressed fear that "should progress in biological research eventually explain all varieties of pro- and antisocial behavior in terms of metabolic and genetic processes...the traditional views of human morality would evaporate as historical curiosities."[29] Such concerns have been based on a rejection of deterministic philosophies. There is a persistent conviction that humans are free to modify their lives; to overcome their biological heritage; to act altruistically when called upon.[30]

Many prominent biologists have taken positions that imply a moral sense. Waddington, for example, said that "the process by which the developing human individual acquires the ability to take part in the sociogenetic transition of information also endows him with a set of conscious purposes of the particular character we recognize as ethical."[31] He was referring to "sociogenetic" information and conscious ethical purposes. (However, he provided no evidence for the existence of such influences.)

This positing of a "ghost in the machine," a principle to be advanced in this text, has been a common theme of moral theorists. Butler contended that "there is a natural principle of benevolence in man; which is in some degree to society what self-love is to the individual.... If there be any affection in human nature, the object and end of which is the good of another, this is itself benevolence."[32] Hume had said much the same thing a century earlier. "There is some benevolence, however small, infused into our bosom; some spark of friendship for humankind; some particle of the dove kneaded into our frame, along with the elements of the wolf and the serpent."[33]

But biologists who accept genetic determinism are to be found in all of the life sciences. And they find a certain amount of support even among those who disagree. Lewontin et al. said: "biological determinism...is not the product of a fringe of crackpots and vulgar popularizers, but some of the core members of the universities and scientific community."[34] Smith, however, expressed the reservation that "at best [biological determinism] could mean that most human beings behave in [certain] ways in [certain] environments."[35] That interpretation should surely take some of the extremism out of such claims.

In point of fact, there is a pertinacious attack on sociobiological claims, amounting in some cases to a total denial of their legitimacy. Kaye took the position that Monod, in representing the view of most reductionist biologists, was as mystical and purpose bound as those he criticized. "The direction of Monod's thought, like that of his fellow pioneers in molecular biology, has not been from molecular biology to metaphysics, 'from biology to ethics,' and from science to values, but precisely the reverse."[36]

On the issue of universalizability, sociobiologists are attacked for providing unnecessarily narrow interpretations. They propose, for example, that incest is virtually always abjured. Psychoanalysts apparently in agreement with the sociobiological notion of genetic dominion, contend that the urge toward incestuous behavior is an innate impulse which is controlled by the dominance of a jealous father. Others have based it on an avoidance of those with whom one has developed close emotional ties on a sibling basis. The sociobiologically inclined make the argument that genetic problems associated with the pairing of similar gametic content, should cause incestuous behavior to "disappear" over time. Thus, the issue of immorality as it relates to incest becomes reduced to something more closely related to ethics.

The reason that people don't practice incest widely, they say, is simply because a particular society, at a particular time and, for idiosyncratic reasons, abjures it. No underlying affective state need be involved. Most social anthropologists believe the term is too abstract a generalization, but they apparently do not deny the possibility that such an impulse exists.[37] They believe that the same kind of research can be applied to other behaviors until the notion of altruism, like other "moral" concepts has been eradicated. Once again, the principle of self-interest—the primacy of the gene—is secured.

Among the milder reactions to the reducing of the influence of social factors was that of Montague, who said that "a besetting sin of sociobiologists is that they are prone to start out with the assumption that they are going to find a hereditary basis for the social behavior they observe, and invariably succeed, by analogy or extrapolation, or misinterpretation, in confirming their anticipated findings.... Universality of a trait does not constitute evidence of genetic determination."[38]

There is, occasionally, muted acceptance of the possibility that sociobiological principles have something to offer. Williams said that "we have no reason to think that there cannot be, in principle, effective and quite strongly constrained sociobiological explanations for these [altruistic] kinds of phenomena. It is not ruled out, a priori, but the constraints on their being so seem to be both tight and secure."[39] Such a position is espoused by a number of humanist biologists. Wolff allows that biology "prepares the structural conditions" that allow for decision making, and thus moral behavior.[40] Biological humanists, however, deny the inevitability of interpretations based on genetic findings.

> The conflicting interpretations of...rigorous scientists would seem to challenge the naive assertions of molecular biologists, while the sense of urgency and the moral passion informing the social and philosophical writings of biologists would seem to challenge the simplistic dismissals by their critics.[41]

Wilson, who in 1975 had said that "the theory of group selection [which he then supported] has taken most of the good will out of altruism,"[42] apparently accepted the existence of a moral sense in 1989. "Innate censors and motivators exist in the brain that deeply and unconsciously affect our ethical premises; from these roots, morality evolved as instinct."[43] Furthermore, he said that as more is learned about our biological processes "we will have to decide how human we wish to remain...because we must consciously choose among the alternative emotional guides we have inherited."[44] Not only does he now appear to accept a moral sense (instinct), he believes people are free to choose between behaviors. He has come a long way from the determinism that sociobiologist principles mandate, and that he once supported.

Culturists argue that the role of altruistic behavior is to make the lives of members of a society more rewarding. They contend that through sharing behavior and the acceptance of responsibility, undesirable traits such as aggression may be contained. As to the claim that behavior is innately selfish, Schwartz voiced the conviction of many psychologists and sociologists that people are not selfish by nature. "The social condition in which people live can make them selfish and greedy."[45]

Several types of argument for the existence of a moral sense that calls for self-sacrificial behavior have been offered. Waddington proposed that morality may be built into the genetic apparatus. Markl disagreed, however, saying that "there is not the slightest evidence for assuming that there are genes for encoding specific moral capacities."[46] Though it is a contention of this text that Markl was wrong in his conclusion, it is true that those who deny the rigorous self-serving model espoused by biologists provide little tangible support for their position.

Many of those who oppose the sociobiologist position do so on the basis of a conviction that humans are significantly different from plants or simpler animals; that an appeal must be made to some "humanizing" property that is unique to homo sapiens (and perhaps even to more primitive members of the class of hominids). This notion of relating the behavior of animals, as well as humans, at least in part to their biological infrastructure has found support among a number of anthropologists. Fox makes an elegant plea for the acceptance of animal/man commonalities, and a rejection of the notion that culture is independent of its biological roots.

His thesis is that among all animals (though especially in humans), genetic or biological endowment limits the kinds of culture that can develop. "The potential for culture," he argues, "lies in the biology of the species. Man has the kinds of culture and societies he has because he is the kind of species that he is."[47] However, he recognizes that "like language, the capacity for specific kinds of behavior is in man, but exactly how this will be manifested will depend on the information fed into the system."[48] The issue of the source of that information is the topic to be addressed in this text.[49]

Perhaps Nagel best exemplifies the confused state of affairs among those that find current, neo-Darwinist biological explanations untenable. He proposes that ethics "is the result of a human capacity to subject innate or conditioned prereflective motivational and behavioral patterns to criticism and revision, and to create new forms of conduct."[50] While such a capacity is undeniably a characteristic of human—as well as much animal—behavior, one may question the connection between innate behavioral patterns and moral or ethical behavior. It is because of the unassailable, as well as unsatisfactory, nature of such "idealist" positions that attention shall be focused on the claims of sociobiologists in this text.

The idea that humans do not act freely, do not make autonomous decisions, and are not responsible for their actions, violates what appears to many to be powerful evidence. Stent offered an impressive challenge:

> As scientists we must regard the person as a material object forming part of the causally determined events of the natural world; but as moralists we must regard the person as an intelligent subject forming part of the world of thought *that is independent of the laws of nature* (italics added).[51]

His statement included a number of rather remarkable claims. He suggested, first, that those who agree with this position are the true moralists, and secondly he offers the egregious assertion that mind as an aspect of "personhood," is independent of biological influence. The acceptance of his thesis would be to define morality in such a way as to eliminate any possibility of tracing it to its biological roots. However, many have agreed. Kummer referred to "the fact that at least the conscious goals of moral norms can be independent of biological ends."[52] He apparently believes in the existence and influence of unconscious norms.

Though respected biologists abjure the claims of creationists, their own insistence that man must be seen as unique, and that he marches to a different (moral) drummer, gives one pause. Scott proposed that "many scientists seem to have a faith in [the spontaneous origin of life which is] as firm as the belief of any religious devotees."[53] Although Mayr—and probably most biologists—argue strenuously that life happened as a contingency, rather than by design, their contention that the moral behavior of humans is of a different class than that of other animals encourages a suspicion that creationists provide a convenient whipping boy. The writings of many eminent biologists reveal that they experience emotional responses that they must assume to arise adventitiously.

In challenging the notion that all evolutionary development centers on genetic alteration, many biologists point to the paucity of knowledge concerning the relationship between genome and phenome. Stebbins contended that "one of the biggest gaps in modern biological knowledge is our almost total ignorance of the mechanisms by which disparate programs of differential gene action are brought about."[54] Stent was quite specific:

> To cover up this conceptual deficit, sociobiologists arbitrarily define the gene as that hereditary unit that natural selection happens to select. But that unit is certainly not what geneticists mean by "gene." The sociobiological "Selfish Gene" is neither selfish, in the context of morality, nor is it a gene in the context of genetics.[55]

It is not certain that geneticists totally disagree with that definition of a gene, nor is it at all clear that the label "selfish" is not an accurate term to apply to the tendency of individuals to act in ways that favor the selection of those genes that are considered most fit, or that comprise the phenotypes which carry them.

Psychological Interpretations

A variety of positions are taken by the members of the psychological community, most, but not all of which, are consistent with the views of those who believe that a moral sense exists. Social learning theorists contended that altruistic behavior is shaped by "social compensation," by the acceptance extended to those who act in a nonselfish manner. To this extent, moral learning is assumed to be the equivalent of any modification of knowledge. Thus, the theory is, in a sense, morally neutral. While their assumption is that reinforcement and other conditioning techniques can be employed to develop positive moral behaviors they have provided no evidence that such a venture should be undertaken. Hull, Spence, and especially Skinner, took the position that individuals "do not need to be rewarded for particular responses in order to develop orderly [morally responsible] behavior."[56]

Whatever one is doing when a "reinforcement" occurs, they say, will be done again under similar circumstances. Under such an interpretation there is no need to attend to a moral sense, or any other mental phenomenon. Fortunately, particularly since the new emphasis on cognitive psychology, such positions are considered today, to represent the voice of a lunatic fringe.

A significant number of psychologists have insisted that behavioral research provides data that support the contention that true altruism exists. Batson proposes that:

> For many years, psychology, including social psychology, has assumed that we are all social egoists, caring exclusively for ourselves. Recent evidence suggests a very different view. It suggests that not only do we care but also that when we feel empathy for others in need, we are capable of caring for them for their sakes and not our own.[57]

There is in this statement, of course, the reservation that caring behavior may occur only when we "feel empathy for others in need." However, Batson and his colleagues have done extensive research that focuses on the behavior of individuals in a wide variety of settings. The welfare of others has been shown to take precedence over that of the self in many instances, even where a degree of pain and suffering must be undergone by the generous behaver.[58]

The psychoanalytic interpretation of the genesis of moral behavior has its roots in postulated biological entities such as a superego, a "censor of human behavior [that] rewards, punishes, recommends, and prohibits."[59] A basic premise of psychoanalytic theorists is that id (pleasure principle) impulses must be countered by the reality principle which, in essence, refers to the objective world in which the mature person must live. Wolff expressed this view in his statement that "the general claim that all behavior is always determined by both genetic and environmental factors is obviously true and cannot be refuted."[60] The genetic factors are assumed to be selfish (id processes), the environmental factors socially administered constraints (the superego).

Psychoanalysts take the position that the individual lives in a cauldron of libidinal and aggressive urges which must be held in check by social regulations. They suggest that children experience a continual conflict between suppressed instinctual urges, and family and other social restrictions placed on the expression of such desires. They learn to be moral in order to obtain rewards and avoid punishment, although "the individual is unaware of the moral demands of the superego which influence his thoughts and behavior."[61] But being "good" in order to be rewarded does not seem in any way to represent moral behavior.

> Why should the boy, punished by his father, learn anything except that he will suffer if he persists in inappropriate behavior? How can such an experience lead to a sense of value or of moral propriety? Although the child can learn that his father associates "goodness" with an act, from what source did the father learn this affective state? And why should the child incorporate the selflessness [that such behavior] implies?

Cognitive-developmentalists compare altruistic behavior with prudent (long range self-serving) action. Kohlberg, a leading proponent of that school, suggests that "both require [the] capacity to delay, [with] preference for the distant greater gratification over the immediate lesser gratification."[63] The distinction is based on the target of the behavior. Here Kohlberg takes a position that reflects the contention of this text. Although most studies of altruistic behavior conclude that "the distinction between justice...and altruistic behavior is enduring,"[64] he argues that "reasoning about justice is required to resolve the basic conflicts between persons in society [and] a single attitude...may be seen as underlying the principle of respect for persons."[65] That "attitude" shall be shown here to be based on a moral sense. Kohlberg is, of course, most interested in the cognitive, or rational process that characterizes various stages of moral development. However, he proposes that "a distinctively moral judgment is a necessary component of an action judged moral, but it need not be sufficient for evaluating the morality of an action or actor."[66]

A principle tenet of Kohlberg's thesis is that moral behavior is associated with positive mental health. "The person at a higher level of moral development," he said, "deals with...problems more effectively or more adaptedly than does a person at a lower level."[67] Further, he claims that social conflicts are best dealt with at the highest levels of moral reasoning. "If you want to make decisions which anyone could agree upon in resolving social conflicts, stage 6 is it."[68]

Studies by Wonderly and Kupfersmid, however, have challenged both claims. They showed that "moral judgment (i. e., cognitive processing of moral data) is not predictive of mental health status."[69] Furthermore, they demonstrated that there is "sparse evidence that the promotion of postconventional [stage 6] decision-making will in any way result in necessarily more 'just' resolutions and/or more desirable behaviors and affective states."[70] The claim that a lack of psychological balance at any stage of moral development is the cause of moving to a higher stage—has also been shown to lack experimental support. "At present there is little empirical support for the contention that disequilibrium is associated with moral stage transition."[71] Thus, little of the Kohlbergian thesis provides experiential or evidential support.

Theological or Creationist Convictions

Joining the ranks of those scientists who take issue with the sociobiological interpretation of altruistic behavior are individuals who represent various religious sects. That view is often considered a "traditionalist," or "far-right-wing" philosophy. Prior to the advent of the scientific revolution, it was assumed by most people, including many philosophers, that individuals are agents of God; that the function of humans (as well as animals) is to procreate; to "people" the world. The unselfish (nonsexual) love of one person for another ("agape") was considered by many to represent the highest form of virtue.

That conviction is shared by billions of individuals today. For such people, very little if any authority or prior significance is vested in the individual except, for example, as a parent controls a child, and in such cases the process is assumed to be guided by advice provided by earthly representatives of an eternal being. Altruistic behavior is an example of the morality that is assumed to be an innate characteristic of all people, and which represents a commandment of God.

For the religious, charity is perhaps the ultimate form of altruism, in that no return is ever expected. Among Jews, for example, "caring for one's fellow man is not merely a generalized moral commandment.... It is spelled out in specific, legally binding obligations which each man must heed."[72] Early Christians claimed that charity "earned merit in heaven and sustained those dear to God, the poor."[73] The church has "helped widows, orphans, and the sick and infirm...and she frequently intervened to protect the lower orders from unusual exploitation, or excessive taxation."[74] This sense of obligation to provide assistance was assumed to descend from a higher power, and, at least in theory, to be based on a sense of responsibility assumed to be ingrained in each individual to attend to the welfare of others, on the basis of the influence of a soul.[75]

With the advent of the "age of reason" it has become appropriate for the educated person to abjure allegory, and to accept the inconsequential nature of life. The result of such a revolution of thought has, by and large, provided for a great improvement in the physical conditions of life. Progress in medical treatment, communication techniques, industrial practices, etc., have made possible a more affluent, healthy, and secure existence for millions. However, with such advances have come, according to the "defenders" of the faiths of various types, a number of unanticipated—often baleful—spin-offs. Most serious, they contend, is scientism, "the belief that the assumptions...of the physical and biological sciences are equally appropriate and essential to all other disciplines, including the humanities and the social sciences."[76]

This myopic view of the method of explaining the nature of life processes, say creationists, provides for a "scientifically correct" interpretation of biological phenomena; a tendency to accept interpretations because of the prestige of leaders of a discipline. They point out that some of the results of the uncritical acceptance of scientific pronouncements have been observed in predictions of global warming, nuclear winter, ozone depletion, deforestation and other threats to human life, most of which have not stood the test of careful scrutiny. Those who are critical of the attempt to reduce human behavior to similar scientific interpretations find sociobiology to represent an unacceptable, extremist, philosophy. They would wonder whether Ghiselin's contention that all who consider themselves altruists are hypocrites doesn't tell more about Ghiselin himself than about those he ridicules.

Psychosocial Models

Although the assertion that all behavior has the ultimate purpose of enhancing the inclusive fitness of the behaver is at best only one of many plausible interpretations of seemingly altruistic behavior, no alternative model proposed to date has made a convincing case for its acceptance. Most, in fact, can be reduced to principles that are suspiciously similar to those of reciprocal altruism. In point of fact, the vast majority of objections to the sociobiological interpretations discussed above are without substance.

An early rebuff to the selfish gene hypothesis was offered in the 1960's, when sociologists challenged the contention that a major reason that people practice beneficent behavior is for possible "social rewards." It was assumed that a public display of assistance to others may lead to reciprocity. Researchers, however, demonstrated that the probability of the occurrence of helpful behavior is negatively correlated with the number of people who are aware that aid is needed. Latane and Rodin suggested that in situations where many people are available, the responsibility is believed to be shared. Each person can assume that others are available to solve the problem. Experiments were performed in which individuals not seen but within hearing range of subjects emitted cries that suggested they were in great distress. In each instance, help was more apt to be offered when the potential "altruist" was alone. The researchers concluded that "there may be safety in numbers, but these experiments suggest that if you are involved in an emergency, the best number of bystanders is one."[77]

More recently, Hoffman and others have suggested that empathy is a major cause of genuine altruistic behavior. Krebs, for example, said that

"What shared genes are to the biology of altruism, empathy is to the psychology of altruism."[78] Hoffman agreed, challenging the contention that empathy is egoistic because it is an aversive state, contending that:

> if a satisfied feeling results from action triggered by empathic distress, then that satisfied feeling cannot be reason enough to describe [it] as egoistic. To do this would eliminate the possibility of altruistic motivation by defining it out of existence.[79]

He pointed out (quite accurately) that the proposition that helping others may be based on "hidden, unconscious, or tacit self-serving motives (e.g., social approval, self esteem)...is just an ad hoc hypothesis, not evidence."[80]

Hoffman was impressed by the fact that members of highly individualistic societies such as America, "usually respond empathically and altruistically to others in distress,"[81] stressing his caveat that "it becomes difficult to accept the prevalent but untested assumption that all seemingly altruistic behavior derives from egoistic motives."[82] He bolstered his argument with the claim that "there is considerable evidence that people of all ages respond empathically to another person in distress."[83] He contended that research data support his interpretation, since "empathic distress has been shown to contribute to helpful behavior and diminish in intensity following such behavior, but [tends to] continue at a high level in its absence."[84]

Genuine altruism, however, cannot be assumed to be manifested by empathic behavior. Empathy has been described by psychologists as representing the recognition of one's self in the resemblance presented by others. It is, in a sense, the experience of seeing one's self in a mirror. The very empathic experience has been described as calling forth such comments as "what if that were happening to me, or to my child." Sociobiologists have no problem with such an interpretation. If individuals behave "altruistically" when they observe another person to be suffering because they see themselves in the individual that they help, selfishness has not been explained away. The individual assisted is simply a reflection of the self.

In the case of Latane and Rodin's research, it may be that in the absence of other people, the potential altruist is more acutely aware of (i.e., feels an empathic relationship with) the individual in distress, and is, in fact, behaving selfishly. If a case is to be made for the existence of genuine altruism, it will have to be based on incontrovertible evidence that people sometimes act in ways that serve the interest of those who are indisputably "others." A condition of such behavior would be that it represent a cost to the behaver unrelated to assertive and/or protective urges.[85]

Simon proposes that the docility of most humans may be the basis for altruism. "Docile persons tend to learn and believe what others in the society want them to learn and believe. Thus, the content of what is learned will not be fully screened for its contribution to personal fitness."[86] Furthermore, he argues that such persons "are more than compensated for altruism by the knowledge and skill they acquire,"[87] which includes the ability to function effectively in social settings. He concludes that "there is no reason to rule out altruism as an important motivation [since it is] wholly compatible with natural selection and is an important determinant of human behavior."[88] His definition of altruism is inappropriate, however, since he argues that "altruists will increase in the population as long as the cost of altruism is outweighed by the social benefits."[89] Once again, anticipated reciprocity is the motive.

Many psychologists, in agreement with Simon, have argued that altruistic behavior is learned, representing no more than conformity to societal demands. Mussen et al., for example, contended that "Altruism... in social interactions is a predominant value for middle class parents, one they are eager to have their children acquire.... Parental stress on achievement and cognitive function makes it clear that these standards include maturity of moral judgment and consideration for others."[90]

Such explications miss the point. Why do "middle class" parents set such standards? Are they designed to attract rewards for their children? The fact that people learn "what others want them to learn," or behave on the basis of "parental stress" is obviously true. However, it provides no evidence that genuine altruistic tendencies exist. People learn to eat jalapeno peppers and to drink tequila, but the basis of their enjoyment of such activities is not simply the fact that they are encouraged to try them! What is learned is that certain experiences have value that is related to innate desires.

Many prisoners incarcerated for long periods of time, and servicemen similarly separated from companions of the opposite sex, often learn to practice homosexual behavior. In spite of the absurd contention of Freudian analysts that such acts reveal inherent homosexual predispositions, the cause of this type of perversion lies in the capacity for learning alternative routes to sexual gratification. The partner in such cases represents a vehicle for the expression of a desire for sex. In the case of altruistic behavior, what people come to recognize is the positive affect associated with activity that assuages a different class of desire; one that initiates a willingness to provide assistance to others at risk to themselves, which represents a manifestation of a moral sense.

The unit of selection

In considering the relevant merit of opposing positions, it is essential to understand the implications of the basic principle of selection. That concept refers to the process that results in the preferential survival and reproduction of specific members of plant and animal species. Neo-Darwinian evolutionist theorists describe the process quite specifically as the natural selection of those who are best adapted, which they define as the process in nature:

> by which organisms best adapted to their environment tend to survive and to transmit their genetic characteristics; differential reproduction of different genetic types within populations, subspecies or species."[91]

This broad set of definitions allows for a variety of interpretations, and biologists have embraced them all.

Genes and individuals. The basis for the development of the sociobiological position, as has been described, lies in their insistence that the unit of biological inheritance is the gene. Dawkins said that "it is its potential immortality that makes a gene a good candidate as the basic unit of natural selection."[92] (However, he made it clear that it is not necessarily the operant gene but its successors that are selected.) Li and Grauer took a similar position. "Although individuals are the entities affected by natural selection and other processes," they said, "it is…genes that change over evolutionary time."[93]

While Dawkins and other sociobiologists have promoted the gene or the phenotype as the target of selection, Mayr, and others have opted for the individual as the critical focus. "The individual, not the gene, must be considered the target of selection,"[94] said Mayr, since it is individuals that are, in fact, the organs of reproduction. Brandon and Burian spelled Mayr's position out, explaining that "those organisms which have a version of [a varying] trait advantageous with respect to the prevailing environment will tend, other things being equal, to be disproportionately represented in the next generation."[95]

The notion of dealing with species as individuals has had a certain amount of support. Hull pointed out that "numerous authors have argued that species are the same sort of thing as genes and organisms—spatiotemporally localized individuals.... If species are conceived as the same sort of things...it is at least possible for them to perform the same functions in the evolutionary process."[96]

Groups. The contention that genes or individuals may be selection units is challenged by many biologists who believe that the population, the species or other entity may in many, if in not all, cases be the appropriate target. Ruse, for example, denied that individuals—even species viewed as such—could represent selection units. He argued, "it is the group, the society, that is the unit of selection,"[97] admitting, however, that "one can only suspect that Darwin's sympathies today would lie with those who push individual [or genetic selection] a very long way."[98]

Peacock extended Ruse's position to deal specifically with species level selection. "The principle of natural selection operates on the species rather than the individual level."[99] Williams (1989) pointed out, however, that although "most evolutionary biologists today believe that group selection must take place...it is a weak force, with little explanatory value for the phenomena that they encounter."[100] Thus a variety of existential entities; species, demes, gene pools, and groups in general have been proposed as the locus of selection. Group selection adherents contend that societies which include altruists among their members will outreproduce those that don't. Dobzhansky's description was prototypical. "With group selection, tribes containing many altruists may well have an advantage over those the members of which fend only for themselves."[101]

Stebbins argued that "two kinds of motivation [have evolved]—a desire for power and a desire for approval."[102] On the surface this seems a plausible thesis. People with such motives would be valuable members of their community. "Their contribution to the group as a whole would increase the group's ability to compete,"[103] and, thus, in all likelihood to enjoy enhanced fitness. Mayr offers potential legitimacy to the group selection hypothesis through the application of a statistical model. He, in fact, denies the primacy of the individual gene. In a widely acclaimed article, *The Unity of the Genotype*, he argued that "genes are not the units of evolution, nor are they as such, the targets of natural selection. Rather, genes are tied together into balanced adaptive complexes, the integrity of which is favored by natural selection."[104]

Wilson agreed, suggesting that the theory of group selection has "provided insights into some of the least understood...qualities of social behavior.... The individual is forced to make imperfect choices based on irreconcilable loyalties—between the 'rights' and 'duties' of self, and those of family, tribe, and other units of selection, each of which evolves its own code of honor."[105] There are sufficient data here for an extensive analysis. First, there is an obvious acceptance of a sense of "loyalty." Beyond that, there is a requirement that one "choose" between self care and the interest of some group. But that is the argument of those who deny genetic and/or individual selectionism.[106] Where does one look for an explanation?

Ruse offered a possible defense of the argument for group selection, though he pointed out its paradoxical nature. Many biologists, he said, define species as individuals. The reason for this interpretation is probably because:

> They have an integration, an internal organization, which binds them into a unified whole, quite unlike a normal class or natural kind.... A class cannot change over time. Species evolve and change over time as, in analogous manner do individual organisms.... But this seems to go flatly against the renewed biological emphasis on individual selection. By stressing the unity of species one is downplaying the fact that every gene—is set against every other organism or gene. There may be cooperation—but it must redound to the benefit of the individual.[107]

Ruse's interpretation has implications for this text. The position to be developed here is that the genetic basis of moral sentiments can be shown to have community wide implications. Consider the question of the target toward which DNA expression must be directed. If it is the interest (fitness) of the gene involved, any action that serves other genetically based phenotypic elements must ultimately be self-enhancing. However, in many instances the element involved is not served. Leukocytes, for example risk extermination in activity that serves the total phenotype.

It is in fact the genome or some collection of genes whose fitness is enhanced in such situations. Why should only the individual gene, rather than its group (e. g., population, or species) be served? Brandon and Burian summarized what they believed was probably Darwin's view and which they assume characterizes the position held by most biologists today.

> What evolves is a population; what selection acts on are the (competing) organisms that make up the population in a given generation; and the reason that selection of this sort is effective is that what reproduces differentially are individuals with traits which are differentially adapted to the environment.[108]

Williams struck a note of caution. While he agreed that groups may be selection units where adaptation is not explainable on the basis of genetic selection, he contended that "if there are no...adaptations, we must conclude that group selection has not been important.... Group selection and biotic adaptation are more onerous principles than genic selection and organic adaptation."[109] Darwin, himself, was so much of an individual selectionist that "his acceptance of group selection for morality seems to have been motivated more by the negative cause of being unable exactly to see how individual selection can cause morality than by the positive cause of thinking that group selection validates itself on its own merits."[110] Thus, he concluded that the issue of fundamental relationships—the ultimate source of genetic knowledge—must be considered in any analysis of altruistic behavior.

Summary

With the development of the New Synthesis in biology, along with the dramatic rise in the acceptance of genetic and molecular biological findings a schism has appeared in the ranks of professionals in the social and biological sciences. A critical issue involves the question of morality, and particularly what appears to be altruistic behavior. Those who identify themselves as sociobiologists, contend that genuine altruism cannot occur, while a number of professionals from all behavioral fields are convinced that a moral sense that guides generous behavior is an inherent aspect of the human psyche.

While the majority of biologists, ethologists, anthropologists and other life scientists, as well as those that represent religious convictions, take one side or the other on the issue of the role played by genetic factors, a number of individuals have suggested that a compromise between opposing positions is possible. Lumsden and Wilson proposed a "gene-culture coevolutionary theory," which they defined as "any change in the epigenetic rules due to shifts in gene frequency or in culturgen frequencies [a term they coined to parallel Dawkins' "meme"] due to the epigenetic rules, or both jointly."[111]

It was their position that both culture and genes are necessarily involved in the evolutionary process. They were concerned with "ways in which genetic and cultural evolution can interact through programs of individual development.... Pure cultural transmission can be sustained," they contended, "only with controls that depend on continuing genetic evolution."[112]

Fox also writes extensively on the wedding of biology and culture. He asks:

> "How did culture get into the wiring? How did the great constructors [gene groups] operate to produce this feature, which, like everything else about man, is not anti-nature or superorganic...or any of the other demagogic fantasy states that science and religion imagine for him?"[113]

Perhaps Masters made the most meaningful proposal, with his suggestion that if human behavior is to be understood, there must be a "willingness to abandon the belief that answers are either/or: Our behavior can be both innate and acquired; both selfish and cooperative."[114] That viewpoint is, of course, not new, but it does make the point that extremist interpretations of altruistic behavior do not provide the best hope for a useful resolution.

Chapter 2 Notes

1. Wilson (1975), p. 562
2. *Ibid.* p. 3
3. Ruse (1988), p. 773
4. The term "genaddicts" is proposed here because of the vehemence with which adherents of the sociobiological movement have made their case.
5. Crick (1966), p. 5
6. Williams (1989), p. 195
7. Dyson (1985), p. 74
8. Kaye (1986), p. 50
9. Ruse (1989), p. 47
10. Plotkin (1988), p. 56
11. Ruse (1989), p. 64
12. Wilson (1975), p. 563
13. *Ibid.* p. 562
14. Ghiselin (1974), p. 274
15. Trivers used as an example the saving of one drowning person by another. "Altruistic behavior can be defined as behavior that benefits another organism not closely related, while being apparently detrimental to the organism performing the behavior, benefit and detriment being defined in terms of contribution to inclusive fitness." (1971, p. 35).
16. Dawkins (1976)
17. Solomon (1980), p. 256
18. Mayr (1982), p. 82
19. *Ibid.* p. 83
20. Hobbes (1962), p. 105
21. *Ibid.* This was a political interpretation of the impetus to altruistic behavior; a subtle acceptance of the principle of reciprocal altruism. Such philosophies written a century before the development of sociobiological theory presaged the conviction of the self-directed basis of behavior.
22. Kaye (1986), p. 3. Monod agreed, adding that science will destroy the "religiosity, scientistic progressivism, belief in the 'natural rights' of man, and utilitarian pragmatism to which so many individuals feel bound," (1971, p. 171). The basis for this shift toward a reductionist stance has been, in part, the result of research that suggests that the gene is the appropriate unit of selection, and thus that selfish behavior is to be expected.
23. Dawkins (1995), p. 85
24. Hume cited in McTaggert & McTaggert (1967), p. 370
25. Mead (1967), p. 374

26. Waddington quoted in Plotkin (1988), p. 6
27. *Ibid.* p. 50. One apparently assesses the implications of an "altruistic" act. What will it mean for me? The sociobiologist position is that any such characteristic as altruism, does not—could not—exist. Such concepts as kinship preference and reciprocal altruism are presumed to account for what appear to be other-directed behaviors. All behavior is based on the hope that the extended favor will be returned—perhaps with interest. Of course, not all sociobiologists are in complete agreement. Hull stated that "sociobiologists do not form a homogeneous group, either conceptually or socially.... They cooperate to the extent necessary to oppose their joint enemies." (1989, p. 289). But neither are their opponents of one mind. "The critics of sociobiology assume that the ideal society would be egalitarian, cooperative and socialistic. To them, the reputed individualistic implications of sociobiology are not in the least alarming," (*Ibid.* p. 277).
28. Wolff (1980), p. 90
29. *Ibid.* pp. 84-85
30. "The major objection [to sociobiologist claims]," said Hull, "is that human beings are conscious, moral agents," (1989, p. 257). Furthermore, said McShea: "The actual findings of the sociobiologists supports neither their assumption that behavior is genetic in higher animals nor their reduction of motivation...to the single selective process that called those feeling into being," (1990, p. 172).
31. Waddington (1971), p. 24
32. Butler (1873), p. 27
33. Hume (1956), p. 271
34. Lewontin, Rose & Kamin (1989), p. 181
35. Smith (1980), p. 29
36. Kaye (1986), p. 76. Trigg was adamant. "We can learn much about ourselves by listening to sociobiologists. We learn nothing about morality," (1983, p. 148). Ruse commented that "claims about altruism and the like––the very heart of human sociobiology—have been highly controversial, calling down derision and scorn from many, particularly left-wing biologists," (1988, p. 65). Alper said that sociobiology is "a reductionist, if not reactionary exercise doomed to futility and marred by ideological presuppositions," (1978, p. 207). It is accused of "giv[ing] pseudo-explanations, thinking up 'just so' stories to explain any and every human phenomenon from an 'adaptionist' standpoint," (*Ibid.* p. 67).
37. Solomon was highly critical. He complained that "social universals...[are] only deviously explained by sociobiological principles when sociocultural principles seem...to do the job specifically and convincingly," (1980,

p. 270). He pointed out that "there are a number of societies where marriages between brothers and sisters are encouraged or arranged," (Ibid. p. 272). Kowalski argued that although the incest taboo seems "at first sight [to be] a quasi-universal feature...this claim is [widely] contested," (1980, p. 237). The problem that must be dealt with is: What do philosophers such as Solomon and Kowalski assume to be the basis on which "socio-cultural principles" are founded?

38. Montague (1980), p. 6. Others were less kind. Midgley referred to Dawkins' text, *The Selfish Gene*, as "cheap, crude, B-feature fatalism," (1980, p. 86), and referred to him as "an uncritical philosophic egoist," (*Ibid.* p. 108). Kaye said that "Wilson's curious anthropomorphizing of genes is...a sanctioning myth for his moral and social prescriptions," (1986, p. 130), adding that "Dawkins' myth of the selfish gene and its hellish creation, is, of course, scientifically false, as well as being morally abhorrent," (*Ibid.* p. 141). On a less dramatic level, Rosenberg referenced Midgley who like so many, he said, believes that genes can't be selfish "because selfishness, and altruism for that matter are motives...which can only be attributed to systems capable of having wants, of calculating consequences, and undertaking actions," (1985, p. 249). But from what source does Midgley suppose such motives are generated?

39. Williams (1980), p. 284
40. Wolff (1980), p. 85.
41. *Ibid.* Albert & Denise argue that moral principles cannot be escaped. Even the most cynical moral opportunists [sociobiologists?] in their recommendation that we act in each case only to promote our best interests, are setting up a principle to govern behavior," (1988, p. 3). Is it possible that the best compromise that can be reached will parallel that of physics in which the apparent wave particle duality of matter has resulted in the acceptance of the ineluctable contribution of an observer to each perceptual experience? Are molecular biologists correct when altruistic behavior is seen from one perspective; wrong when viewed from another?
42. Wilson (1975), p. 120
43. _____. (1989), p. 243. The "innate censors" and "motivators" that Wilson refers to are relevant to the thesis to be developed in Chapter 4.
44. *Ibid.* Notice the "emotional guides" that he assumes exist.
45. Schwartz (1986), p. 310. This, of course is not necessarily a refutation of sociobiological principles, which may easily be expanded to embrace the notion that societies may alter the manner and the extent to which selfishness is manifested. Williams contended that people who make friends and act in socially appropriate ways have an evolutionary advantage. "I imag-

ine that this evolutionary factor has increased man's capacity for altruism and compassion and has tempered his ethically less acceptable heritage of sexual and predatory aggressiveness," (1984, p. 53).
46. Markl (1980), p. 219
47. Fox, R. (1989), p. 20
48. *Ibid.*
49. Fox added that, "The human is like a computer...in a state of readiness—at various points in the life cycle—to process certain kinds of information," (1989, p. 23). Thus, culture will vary with environmental opportunity, but only within biologically prescribed limits. This principle, he proposed, is quite the same for any species capable of developing a culture.

Solomon expresses concern that sociobiologists explain altruistic behavior "in an often ad hoc and speculative way when cultural explanations...are readily available," (1980), p. 271. The approach that they employ, he said, cannot be considered scientific, since it is based on almost precisely the same principles that activate the creationists.
50. Nagel (1980), p. 205
51. Stent (1980), p. 15
52. Kummer (1980), p. 32
53. Scott (1986), p. 83
54. Stebbins (1982), p. 158
55. Stent (1980), p. 12
56. Marx & Hillix (1979), p. 289
57. Batson (1990), p. 336
58. One may ask a related question regarding the concern that is shown by so many animals rights activists when a species of animal is threatened with extinction. They make many efforts, including considerable self-sacrifice, in their effort to have legislation passed which will protect such species. Note that it is the species rather than the individual that is preserved. If those cows, chicken, pigs and other animals that are slaughtered for food and other purposes were to be at risk of extinction, efforts would be made to prohibit their destruction. Would the spotted owl and other animals now being protected lose their favored place should their be a significant increase in their populations, and their flesh be discovered to be succulent, as well as nutritious? In such cases, what profit to those that fight for such legislation can be discerned? It would appear that neither the kin selection process, nor the principle of reciprocal altruism is involved.
59. Wonderly (1991), p. 303
60. Wolff (1980), p. 89
61. Weiss (1960), p. 34

62. Wonderly (1991), p. 333
63. Kohlberg (1987), p. 260
64. *Ibid.* p. 270
65. *Ibid.* p. 293
66. *Ibid.* p. 272. Thus, while allowing for the necessity of the influence of a moral sense, Kohlberg cannot concede that one's appeal to such a sense automatically determines whether an action is, or is not, moral. He is in fact critical of Kant for believing that "intuition of rational conscience [is] inborn in all human beings in all cultures and [does] not depend on experience for its development," (1987, p. 261). Kohlberg, of course, believes that moral judgment develops through a series of "stages," and that there is a consistency in the maturation of such stages among cultures. These judgments "are judgments of value, not of fact.... They are prescriptive," (Colby & Kohlberg, 1987, p. 10). But prescriptions imply oughts or shoulds.
67. Kohlberg & Mayer (1972), p. 480
68. Kohlberg (1971), p. 216
69. Wonderly & Kupfersmid (1979), p. 115
70. _____. (1980), p. 626
71. _____. (1978), p. 303
72. Herzberg (1962), p. 103
73. Fox R. L. (1989), p. 323
74. Durant (1950), p. 78
75. Henry Adams made the point (in 1860) that "only about two hundred and fifty years ago the common sense of mankind, supported by the authority of revealed religion, affirmed the undoubted and self-evident fact that the sun moved round the earth," (Stevenson, 1959, p. 351). Just such convictions are held today as regards the preeminence of God, and the absolute mandate to act in accordance with His will—which includes the obligation to respect at least some other individuals. The limitation of that obligation in many instances to a "preselected" few may be the principle cause of the diminution of influence that the church commands, particularly as it bears on scientific analysts.
76. *The Random House Dictionary of the English Language* (1987), p. 1716
77. Latane & Rodin (1969), p. 201
78. Krebs (1987), p. 104
79. Hoffman (1981), p. 134
80. *Ibid.* p. 125
81. *Ibid.* p. 135
82. *Ibid.*

83. *Ibid.* p. 128
84. *Ibid.* p. 130
85. All behavior will be shown in Chapter 4 to involve costs. The distinction to be made is that self-assertive and self-protective behavior are self-oriented, while the goal of transcendent behavior is the welfare of others, with altruistic behavior being instrumental and, thus, a cost.
86. Simon (1990), p. 1665
87. *Ibid.* p. 1667. He added that inculcating guilt and shame also "serve most people as strong motivation for accepting social norms," (*Ibid.*).
88. *Ibid.* p. 1668
89. Simon referenced in Guisinger & Blatt (1994), p. 106
90. Mussen et al. (1970), p. 192
91. Barnhart (1986), p. 427
92. Dawkins (1976), p. 38
93. Li & Grauer, (1991), p. xiv
94. Mayr (1988), p. 101
95. Brandon & Burian (1984), p. xi
96. Hull (1988), p. 24. Rosenberg summarized the views of Hull, Ghiselin and others. "Each species is an individual, spatio-temporally restricted (although scattered) particular object, whose members are its parts and its components, not its instances," (1985, p. 204). This interpretation provides a significant component of the infrastructure of the model to be described in this text.
97. Ruse (1984). A biological group was defined by Uyenoyama and Feldman as "the smallest collection of individuals within a population defined [in such a way] that the genotypic fitness calculated within each group is not a ...function of the composition of any other group." (1980, p. 395).
98. *Ibid.* p. 23. Hull agreed with Ruse, saying that although a group may be essentially heterogeneous "it might still function as a unit of selection in biological evolution," (1989, p. 249). And Wilson, in speaking of primitive forms of life, made the point that "the primary goal of individual colonial invertebrates and social insects is the optimization of group structure," (1975, p. 243). This certainly represents an example of group selection. However, at another point he said that, "Although the social behavior of the lower invertebrates and insects has evolved mostly through group selection...the social behavior of the vertebrates has been evolved mostly through individual selection," (*Ibid.* p. 381). Thus, he embraces all possible interpretations.
99. Peacock (1976), p. 21
100. Williams (1989), p. 196

101. Dobzhansky (1970), p. 427
102. Stebbins (1982), p. 401
103. *Ibid*. p. 402
104. Mayr (1984), p. 82. He contended that group selection "is valid only for groups with a fitness value that is greater than the arithmetic mean of the fitness values of the individuals of which it is composed," (1991, p. 157), adding that in addition to a group composed essentially of family members, there are "groups of nonrelatives which practice social facilitation or various forms of mutual help," (*Ibid*.). Unfortunately such data provide no hint about why such "mutual help" should be extended. Hull made the same point, saying that the genotype "seethes with activity; genes turning on and off, introns being snipped out, other segments moving from place to place in the genome, and so on," (1988, p. 23). For both men, gene groups or complexes are the appropriate units of selection.
105. Wilson (1975), p. 129
106. Goldschmidt said that "moral behavior in humans is positively reinforced by...the subjects 'acceptance' in his society, " (Goldschmidt quoted in Kummer, 1980, p. 40). Although the term "moral" was used, Goldschmidt was obviously explaining such behavior in terms of anticipated rewards, which assumes either a sense of responsibility in others, or a manipulative technique employed by the one who acts "altruistically." Rheingold and Hay, in their studies with children contended that "sharing behaviors develop because of a 'rich social response' to such behavior," (1980, p. 102). Jenner stated that "allegiance may depend on some innate human bonding," (1980, p. 138). In each instance pay-offs or rewards are implied and the selection unit need not be the social group involved.
107. Ruse (1988), pp. 55-56
108. Brandon & Burian (1984), p. 167
109. Williams (1984), p. 67. Arnold & Fistrup added that when "causation" is under consideration, the critical question is whether "differences in mean individual fitness among different groups are merely an artifact of individual level selection and segregation by character value, or ...represent evidence of group treatment effect on individual fitness," (1984, p. 306).
110. Ruse (1984), p. 23
111. Lumsden & Wilson (1981), p. 11. Dawkins coined the term "meme" to describe a unit of cultural transmission. In a sense, a notion of memory.
112. *Ibid*.
113. Fox, R. (1989), p. 34
114. Masters (1989), p. 1

Chapter Three

Altruism and the Motivational Process I

Existential Principles

> *While (social scientists) are fiddling, Rome is burning, and someone has to attend to the fire rather than debate whether we can ever have sure and objective, non-relative knowledge of the fire's existence.*
>
> Robin Fox

In this section, an interpretation of altruistic action that is consistent with the findings of geneticists and molecular biologists shall be developed. Such behavior can be shown to make a positive contribution to emotional health, since it represents the meeting of a significant human desire—the urge to belong; to find meaning and purpose; to share; to give; to love. Several interdependent factors will be considered. These include a model of life and the motivational process, based on principles of:

- evolution and the adaptive process
- the holonic/holarchic theory; subservience of parts to wholes
- the nature of information, knowledge, and communication
- the status of genes, DNA, RNA, optimons, interactors, etc.
- systems concepts, and

- the manifestation of these tenets in the behavior of organisms at every stage of life from the simplest bacterium to the complexity of human existence.

Evolution

Any effort to address the problem of the genesis of moral behavior and the role of genetic influence on its occurrence must face still unsettled questions among biologists and professionals from related fields regarding the status of the theory of evolution which has become involved in many scientific endeavors. Levins and Lewontin state that:

> The ideology of evolution...has permeated all the natural and social sciences, including anthropology, biology, cosmology, linguistics, sociology, and thermodynamics.... It is a world view that encompasses the hierarchically related concepts of change, order, direction, progress, and perfectability.[1]

The validity of the theory of evolution, since it contends that only the *fittest* survive, is a necessary condition for the acceptance of the sociobiologist's claim, since they apply a fitness test to all behavior. It is not a sufficient condition, however, because of the problem of determining which entity carries the fitness designation. Sociobiologists claim that it is the gene. The contention of this text is that it is the gene *pool*.

Since ethical principles and practices have evolved it would seem logical to assume that such development must have been based on the same type of processes that account for other human characteristics. It behooves biologists and anthropologists, as well as social scientists, to pursue an understanding of *all* of the motives and emotions of humans as altruistic or other self-effacing behaviors are contemplated.[2]

In the effort to understand the source of life, the synthetic theory, or "new synthesis" which is the name applied to the wedding of Darwinian evolutionary concepts, Mendelian genetics, Weismann's "barrier" and molecular biology in general is most widely accepted. Dobzhansky said "the resulting synthesis is truly a biological, not only a genetical or ecological, or paleontological, theory."[3] Fox offered a simple definition. Evolution, he said, is:

> a genetic process whereby the offspring of an organism show some variation of type, that causes them to be selected, by favorable interac-

tion with the environment, to survive and reproduce and give rise to new varieties.... The basic notion is that variability is essentially random and the selection mechanism simply discards all those variations that do not work out as well as the original unvaried stock or as well as a few favorable variants.[4]

However, a number of professionals in biology and related fields have remained unconvinced that the theory in its present form provides even an approximation of an accurate picture.[5] In spite of its acceptance by the vast majority of biologists, they contend that the synthetic theory suffers from a serious defect—life scientists are in utter ignorance of the characteristics of the earliest forms of life. One of the most direct attacks on the "new synthesis" was provided by Ho, who, commenting on the "recent explosion in molecular genetics research,"[6] claimed:

Forever exorcised from our...consciousness is any remaining illusion of development as a genetic programme involving the readout of the DNA "master" tape by the cellular "slave" machinery. On the contrary, it is the cellular machinery which imposes control over the genes.[7]

These are significant challenges to the status quo, and have understandably called forth rebuttal. As is the case with all such issues, writers on each side have made extreme claims.[8] Mayr, represents the conservative position in contending that "all adaptive evolutionary change is due to the directing force of natural selection."[9] He and many others cling to the conviction that *random* gene pairing and *adventitious* mutational occurrences are the principle elements of the evolutionary process. But dissension has persisted. Freedman argued that "this notion [of random mutation] stems from the sheer fantasy that 'in the beginning' there was random activity out of which emerged organization. In fact random mutation has never been observed."[10] Dobzhansky agreed, making the point that "selection acts as an *'engineer'* that 'arranges' and 'constructs' so called chance mutations into *"adaptively coherent* patterns."[11]

Mayr responded vehemently. "Natural selection...rejects any and all determinism in the organic world.... [It] is utterly opportunistic.... It starts, so to speak, from scratch in every generation."[12] "Selection," he said, "is probabilistic, not deterministic."[13] And, "all adaptive evolutionary change is due to the directing force [sic] of natural selection."[14] This totally contingent view of evolution was emphasized by Day, who said that "life is just something that happens.... It is a relationship between matter and energy that occurs spontaneously when the conditions are within narrow limits, and it evolves through stages."[15]

Mayr argues that the synthetic theory is not challenged by alternate views on such subjects as randomness of mutational occurrences at the earliest stage of evolution since "from Darwin on to the 1970's the individual as a whole was considered the target of selection for the organismic biologist and therefore, recombination and the structure of the genotype as a whole were viewed as being far more important for evolution than mutational events at individual loci."[16] The disagreement persists. Li and Grauer contend that "at the molecular level the majority of evolutionary changes...are caused neither by positive selection of advantageous alleles nor by balancing selection, but by random drift of mutant alleles that are selectively neutral or nearly so."[17]

The debate involves the issue of what role non-deleterious mutations play, with conservatives insisting that most mutations are either positive or disappear very rapidly from the genomic makeup. A significant number of biologists, in fact, are convinced that explanations of evolution through the adaptations of "advantageous alleles" exposes a problem inherent in Neo-Darwinist theory. Sinha, for example, asks: "Why, if an organism is 'adapted' in the sense of inclusive fitness, should evolution occur at all?"[18]

A daunting problem is created by biologists' efforts to maintain a scientifically respectable deterministic position while discussing the novelty that develops out of efforts to improve fitness. Rosen points out that "the most awkward aspect of the evolutionist position is that it stands uncomfortably between the hard-rock causalist, deterministic view of reductionist physics and the neo-Darwinist view of an emerging process.... [Modern biologists] believe wholeheartedly that everything about organisms is shaped by essentially *historical, accidental* factors, which are inherently unpredictable and to which no universal principles can apply."[19]

"What is relinquished here," says Rosen, "is nothing less than any shred of logical necessity in biology, and with it, any capacity to actually understand."[20] This is a serious charge. How can the contingencies of environmental happenstance play a role, if the infrastructure is, in principle, precisely identical to that of the nonliving? It should be obvious that any effort to deal with the problem of morality and the altruistic behavior that it encourages must be affected by one's conviction regarding the relative merit of these conflicting viewpoints. It is in this context that the claims of currently popular interpretations of altruistic behavior must be considered; the process of *adaptation* must be understood.

Adaptation

From the time that the concept of adaptation, which, in one sense means to act deliberately (as when one is "willing" to adapt to a particular set of circumstances) became involved, considerable confusion has been generated. Wills speaks of the "adaptive explosion" that ushered in the Cambrian epoch, Mayr and Leakey of "adaptive radiation," and group selectionists speak of adaptation as occurring in the interest of what is "good for the species." In each case an assumption can be (and has been) drawn that some purpose or direction is involved. For some individuals, (*orthogeneticists*), alterations in a gene pool occur in order to benefit someone or some group. A predetermined course is believed to have been set. Most biologists, however, deny that any controlling factor is involved.

A seriously misleading factor in the effort to understand the role of altruism is the obviously unintended implication of sociobiological adherents that choice, and thus lineage enhancing directedness, is involved in the genetic process. Dawkins speaks of the fact that the gene "has the extraordinary property of *being able to* (italics added) create copies of itself."[21] In discussing the copying process he points out that "it is not perfect. *Mistakes* (italics added) will happen."[22] Mayr quotes Weismann, a panselectionist himself, as saying "no device in nature is absolutely perfect."[23] Ridley says "the traits of the organism are well designed; they are…adaptations."[24] And Stebbins adds, "Different individuals or populations adopt different strategies in response to their environments."[25]

In each instance the author's intentions are to describe what is understood to be a purely mechanical process. However, in using terms such as "mistake," "perfect," and what is more misleading, the expression "being able to," one might assume that a DNA sequence may *choose* not to replicate. This may represent no more than a semantic issue. But some individuals have specifically proposed directed adaptation. Ruse said that "parts of organisms are directed towards an end, namely that of survival and reproduction. In short organisms have adaptations."[26] And Mayo added that "sexual selection…is a very convincing example of apparently directed evolutionary change, at an accelerating rate, determined by a well understood and unexceptional genetic mechanism."[27]

The fact is that no mindless entity ever deliberately adapts. What appears to be the result of an adaptive process is no more than the proliferation of those individuals or gene pools whose characteristics are best suited to an environmental niche as compared with those of possible competitors. At no step in the process are options considered. To the extent that genetic change occurs—and it does so only at the level of the gene pool—the term "can" when it means "is able to" or "has the capacity to" must be replaced with the term "must" when appropriate conditions are met.[28]

Adaptation is recognized by biologists as being a relative function. The persistence in a gene pool of any change, whether brought about by natural or sexual selection, will depend, at least in part, on existing environmental conditions. This adds credence to the principle that adaptations are not subject to choice. Mayr specifically denies the element of choice in the adaptive process in his criticism of Lamarck's belief in "transformational" evolution, saying that such a viewpoint assumes that evolution designates "a completely gradual process, a change due to a trend toward perfection or adjustment to the environment."[29] But is he quite unequivocal in his refutation of that position? He states at another point that "evolution is not *necessarily* (italics added) progressive; it is an opportunistic response to the moment; hence it is *unpredictable*."[30] In any case, altruistic, or other-directed behavior can be shown to represent an adaptation that serves the interest of a gene pool.

The obvious problem with accepting the notion of purposive adaptation is that a particular option is assumed to be favored because of its relation to some goal. However, nucleic and amino acids, the components of DNA, are simply chemical compounds. In transporting knowledge they do so without regard for its effect on any level of life. When this constraint on the freedom of action that characterizes inanimate entities is appreciated it should be clear that altruistic behavior will have to be explained in some other way than through appeal to the argument that people are of a "higher order" than lower animals, or that human beings must be regarded as "forming part of the world of thought that is independent of the laws of nature," and thus to some extent free of the rigid response requirements of chemical interaction. (The emergence of an "altruistic urge" is just as contingent as is every other behavioral determinant.)

Adaptation is described as fortuitous in order to stress the fact that what goes on in the biological world is influenced only by actions that inadvertently serve the survival and growth interests of some existent. Although this may seem to support the contentions of sociobiologists, the evidence of altruistic behavior in humans as well as in other animal species provides a strong argument for group selection; for the legitimacy of an entity or entities to which individual phenotypes are subordinate. Gene pools shall be identified here as having an influence on the characteristics of genes available to each generation.

What role does the gene play in the adaptive process? How can it be assumed to function in its own interest? The zygotic gene is, among other things, an admixture of the contribution of male and female gametes. The fitness, and thus the adaptive advantage, of any individual that emerges is a product of the separate fitness of each parent. It may represent the mating of one highly fit and one relatively unfit individual, or two superior—or inferior—instances of that gene pool. Adaptation, then, for genes as well as for species, is a gratuitous event. The adaptive process more accurately, in fact, relates to characteristics of phenotypes. "It is individuals upon which selection acts: Individuals survive and reproduce or they die."[31] But what is an "individual." How shall the nature of existence be described?

The Holon

No existent can be completely described by analysis alone. To appreciate the meaning of an entity, its relationship to something more inclusive (some "whole") must be taken into account. An entity bears the designation whole, only relative to its parts or instances. The term *part* is appropriate when the individuals comprising the whole are dissimilar. A kidney is a part of an animal. A fuel pump is part of an automobile. The term *instance* is suitable when referring to individuals in situations in which the whole refers to a collection of similar entities. It is appropriate when the entities are not physically connected. Instances need not be identical. The instances that comprise a strawberry patch may be plants of considerable range of size and even of type. The essential qualification is that they can be identified for some purpose as strawberry plants—as instances of that whole.

Although the relationship between parts and wholes is commonly accepted, there is considerable disagreement on the chronology of the relationship. Do parts create wholes as a result of their interactions? Do wholes control the behavior of their parts? Or are the contributions of each aspect essentially the same? Such questions are obviously critical to an understanding of human behavior.

Some entities are understood to have parts or instances but apparently are not necessarily considered to *be* parts or instances. But this is an inaccurate representation. Wholes are no more necessarily the outcome or result of the actions of parts than parts are the product of their wholes. Catabolic activity, which provides for the break*down* of complex molecules, is as legitimately the end product of a directional process as is the build*up* of the heat and energy produced by metabolism. Bohm's suggestion, regarding inanimate objects is a typical example of the directional interpretation.

> Consider the structure of a house. One begins with the bricks, which are similar in size and shape but different in position and orientation. The similarity of these differences of the bricks leads to the order of the wall. The wall in turn becomes an element of a higher order, in such a way that the similar differences in the walls make the rooms. Likewise, the similar differences of the rooms make the house, those of the houses the streets, those of the streets the city, and so forth.[32]

The comparison is appropriate. However, the "wall" he references is both comprised of bricks, and is an element of the room *at the same time*. It cannot be completely explained by either aspect alone. The wall is *defined*, in fact, in its relation to the room and is *manifested* in the bricks with which it is built. Bohm's description implies that growth in every instance involves moving from the smaller to the larger; from today until tomorrow; from the simple to the complex. But no such existential vector exists in the inorganic world. Entropy and negentropy are characteristics of entities that are dependent on the vantage point of an observer. Fusion and fission occur sequentially. At the level of mind, this periodicity is reflected in the description of an existent at any moment, with mental dissection and assembling essential to its identification.

Koestler coined a term to be adopted here, which provides an interpretation of existence which is hierarchical, while providing for the relative independence of parts.

> To talk of sub-wholes... is awkward and tedious. It seems preferable to coin a new term to designate those nodes on the hierarchic tree which behave partly as wholes or wholly as parts, according to the way you look at them. The term I would propose is "holon" from the Greek *holos* = whole, with the suffix *on* which, as in prot*on* or neutr*on*, suggests a particle or part.[33]

This conceptualization provides a pragmatic alternative. The term *holon* describes *all* existents at *all* levels, since they are neither whole nor part in any absolute sense, but are always an aggregate of parts or individuals in one sense and parts or instances of larger wholes in another. It would perhaps be more accurate to consider the term as an adjective, and to define existents as *holonic*. (This is not, however, to suggest that existents have a holonic *parameter* such as that of extension or color. Partness and wholeness, are relational terms—no entity is endowed exclusively with either characteristic.)

To exist—to be capable of having parts—is by definition to belong at the same time to some greater whole in order to be meaningful. A nail is part of a finger. However, the finger is defined as part of a hand, which in turn is part of a body, etc. On any occasion that an entity functions subordinately, (i.e., as a part or instance), it exhibits characteristics peculiar to that role. In its whole aspect, it functions in a different manner. Some wholes are, of course, extremely transitory aggregations, while others are relatively permanent. Viewed as a whole, the holon is fully fixed; totally determined. Consider a golf ball. As such (the whole aspect) it is certainly not a butterfly or a chandelier. It is uniquely a golf ball (with its roundness also fixed). When its parts are considered, such characteristics as color, number, size, etc. are free of the constraints of being associated with a golf ball. They could—and do—characterize many other entities.

Directional Holons—The Phenomenon of Life

Life shall be depicted here in such a way as to both account for self-oriented activity and to legitimize altruistic behavior, while maintaining most Neo-Darwinian principles. Terms like morality and reason, egoism and altruism, are relevant to only one manifestation of all that exists in the universe. But what sets that form of existence apart from all else? In what way is it unique?

In spite of what may seem obvious distinctions, consensus on the meaning of life has been difficult to achieve. Every authority proposes a definition that focuses on some predisposing conviction. Philosophers of biological science interpret life in terms of its evolutionary characteristics. Scott proposes that life is a form of matter which has "the ability to evolve in the biological sense of the word—particularly the ability of a population of replicators to gradually change in structure."[34] Bernal said that *"life is a partial, continuous, progressive, multiform and conditionally interactive, self-realization of the potentialities of atomic electron states."*[35] Such definitions reduce life to no more than a special case of atomic existence. Bohm provided a typical sociobiological interpretation:

> What is *basic to all* life is a genetic process, in which changes in the genotype are *always* fortuitously related to the experiences of the phe-

notype and in which these changes will survive *only* if they are favorable to continued propagation of offspring in the existing environment.[36]

In all instances biological knowledge is based on an admixture of that gleaned from genetic studies and inferences based on evolutionist theory. But each interpretation limits the uniqueness of life to certain structures or functional characteristics. None meets the conditions necessary to call for the emergence of a moral or rational sense. "Genetic processes," "molecular hereditary storage," etc., provide no information that distinguish the living in terms of the relationship between individuals and societies, between "persons" and "peoples." A focus on the interaction of various levels of living existence is designed to meet that need.

Life may be characterized as that form of chemico-physical matter that manifests entropy resistance and continued existence across time through the interaction of cell and cell group, genotype and phenotype, species and genus, biomass and ecosystem. It began at one or more places several billion years ago, and every past, present, and future manifestation represents only the temporal extension of that origin through the operation of the genetic process. Life in neither egg, fetus, nor neonate represents any more than a step in the progression. In each instance the focused entity is definable only in terms of its relationship to other segments of this intricate network. The moral sense, and thus altruistic behavior, provide evidence of this peculiarity with phenotypes subject to the survival and reproductive needs of evolutionarily stable gene pools.

Living entities, like all that exists, are holonic in nature. Every level of life is defined in terms of more comprehensive existents and is manifested in its instances and/or parts. To "be alive" is to possess metabolic potential that under appropriate circumstances can be actualized. In a strict sense the individual is not an instance of life but of an extant species or gene pool. The temporal factor, the progenitors of contemporary genomes must, therefore, also be considered. In the case of life, the novelty of combinations of organic elements provides for the increasing perseverance of the whole. This represents the creative aspect of existence, with creativity in living beings appropriately viewed as a phenotypic expression of potential for the evolution of gene pools and species.

No living creature can be thoroughly described without considering the community—the deme, population, species, gene pool, or other entity to which it "belongs."[37] There are, however, significant differences between organic and inorganic entities. Unlike nonlife forms, which are subject to all of the elements with no pattern except decomposition or temporary combination with similar senseless structures, living existents exert control over the environment, and over their constituent parts and/or instances. Beyond this, living forms manifest a dynamic survival pattern. This is accomplished in hierarchical order with each ascendant level having an existence which supersedes those elements that manifest it, and derives meaning from some whole it represents. The activity displayed by an individual is, in part, determined by a growth vector of some gene pool of which it is an expression. The process is revealed in the evolutionary continuum, being thus a *homeorhetic* occurrence.[38]

In the interest of developing a theory regarding the existence of genuine altruism, the focus here will be on the relationship between individual "instances" and the gene pool. Gene pools are selected because it is possible to demonstrate that evolutionary stability as a characteristic of human life is exhibited at that level. There is a mutual existential entailment between parts, instances, and wholes, which is manifested in the relationship between individuals and other members of their species. This reciprocal transaction identifies such aggregations as *systems*, a concept that represents a critical factor in the evolution of increasingly segregated gene pools.

Systems

General systems theorists account for life in thermodynamic terms through the concept of the open system, substituting the notion of *feedback*, for purpose. Each system may be viewed as being a potential element of any number of larger systems. In systemic organizations, elements functioning as parts retain their integrity over shorter periods than does the system itself. Systems endure beyond their specific manifestations. When an organism, a population, or a gene pool is considered in its systemic aspect, relational concepts are essential. *Organic* systems have the distinctive characteristic that parts are subordinate to the totality, as stated above. This characteristic—*directionality of function*—is found only in living systems.

The significance of that peculiarity is central to the contention that altruistic behavior is an inherent aspect of the behavior of organisms. Living creatures are instances of more comprehensive existents. They play a role in the systems of which they are components, having the ultimate function of enhancing such systems. Individuals are agents, not only of their own lineage, but of their demes, populations, gene pools, or species. A living entity, viewed in isolation—only as a part—loses its function, or significance; it does not—cannot—participate in the activity of the system from which it is excised.

To completely characterize a system, parts, individuals, and the whole to which they are related must all be identified. Humans as individuals, and *homo sapiens* as a species, exist in a dynamic relationship which may be defined as systemic. The misinterpretation of this characteristic is perhaps the most serious impediment to understanding the nature of a system, *which has the capacity for bridging several holonic levels of existence*. As an entity, a system must be understood to be comprised of both parts and/or instances and some whole, and its function at each level must include all of its constituents.

This characteristic of systems—their seeming evanescence—can be deceptive. A system is the collective existent of which each individual is a component. In the case of humans, for example, when reference is made to serving a gene pool, that entity is a system of which the individual is an instance; the gene pool is the self extended. When a leukocyte is destroyed in the course of its immunological function, it is serving the organism of which it is a part. Similarly, the nucleus, which is a part of a cell serves the totality in which it is located; its more inclusive self. Consider the behavior of an athlete where some sacrifice to the team is called for. The team is not something *there*. It is *here*; it is *now; it is its members*.[39]

Understanding systemic relationships is most apt to cause confusion in the case of *instances* where the relationship is often obscure. Parts have a more obvious systemic relationship to some whole. Furthermore, since instances are in some ways related systemically, and in others non-systemically (as in the case of a nucleus and the class of nuclei), it is possible to ignore the systemic relationship where it does exist. It is easy to confuse "people" in the systemic sense (where genetic influence is involved in their actions), with "people" in the collective (non-systemic) sense that people are, for example, sentient beings.

Inorganic systems are those in which relationships are described in the language of chemistry and physics, and where the dynamism is a force field. Parts and wholes are of equal significance and neither can claim prior existential value except through reference to some external source. In the case of *organic* systems, the interests of parts are subordinate to those of the whole. To some extent the independent or "free" part functions to enhance the (determinate) whole. Such systems represent all forms of life at the levels of ecosystem, species, population, gene pool, individual, organ, cell, organelle, etc. Seed bearing and the reproductive practice in plants are clear evidence of interaction between levels of organic existence. At the level of animal life, sacrificial or altruistic activity provides a more tangible clue to the existence of a compelling and dynamic part/whole relationship.

The genome/gene pool relationship is recognized as a system because of the dynamic relationship between the individuals and the whole in spite of the fact that a gene pool as an existent undergoes constant alteration. One of the most significant characteristics of living beings is, in fact, the systemic organization of individuals, whether as members of relatively well defined populations or of more ambiguously integrated species. Living creatures—phenotypic entities—have the capacity to mitigate the impact of environmental caprice. While nonliving entities are completely subject to the influence of chemico-physical forces, living holons challenge such perils. Although inorganic matter drifts inevitably toward a state of entropy, living beings consistently overcome decay as upper echelons of hierarchical organizations move to higher levels of insulation. A stone, plagued by the incessant pressure of natural environmental occurrences, ultimately loses its identity as its elements degenerate into a less ordered state. Living systems, on the other hand, approach states of extreme improbability. This is accomplished through several processes.

First, living entities, being dissipative structures, maintain their equilibrium through the capture and metabolism of energy sources. Secondly, the program of replication and reproduction causes gene pools and other aggregations of individuals to extend their duration beyond that of specific parts and instances. The phenotypes or individuals that arise out of the amalgamation of genetic and environmental factors represent *components* (parts) of such living systems. Their function (*not their purpose*) is to perform in ways that enhance the gene pool, or species to which they belong.

The preservative characteristic of complex systems represents a challenge to the claims of sociobiologists. Although increasing complexity characterizes larger systems a grave objection can be raised regarding the view that this could occur as an attempt at enhanced survival by individual phenotypes. If life developed according to strict sociobiological principles the relationship between systemic instances (individuals) and wholes (gene pools) would be reversed. The significance of each individual would exceed that of the group in which it participated. The purpose of all behavior would be to enhance the individual and its kin line at whatever risk—or cost—to the evolutionary stability of a gene pool. The gene pool would be relatively insignificant.

The Holarchy and "Independent" Holons

Since life is systemic, it carries out its unique scheme based on a program of differentiation and specialization which insulates higher levels through more efficient function at lower levels; through a systems approach. Koestler proposed that the term *holarchy* be used in place of the term hierarchy which does not reflect the independence that characterizes the function of lower level echelons of organic systems. While hierarchy "conveys the impression of a rigid, authoritarian structure [a holarchy] consists of autonomous, self-governing holons with varying degrees of flexibility."[40] This interpretation identifies the freedom in parts and instances which is obviously advantageous in the pursuit of survival and growth oriented behaviors. It describes humans not only in their determinate whole existence, but as possessing the greatest potential for focusing on their role as instances; as considering personal interests of first priority.

Organic beings are less robust than inorganic entities. Cells that make up the body are more delicate than comparable inorganic configurations. The individuals that comprise a species are extremely fragile as compared to granite, for example, and a species certainly exists more tenuously than an entity such as a solar system. This frailty is compensated for, however, by docility or capacity for modification. To a considerable extent, parts and instances are "free" of total control—are holarchic in nature. That principle may, thus, appear to legitimize the priority of individuals. However, the holarchic relationship extends freedom to lower status levels only because such independence has the potential for serving the gene pool.

At each level of existence, from the lowest discernible living form each individual ordinarily acts partly to enhance the survival of some whole to which it belongs, directly or mediately. In most instances the whole to which the individual is related exerts systemic control (e.g., as in genetically based reproductive capacity). When the need for survival and growth arises individuals at lower levels are often forfeited—in some cases through self-sacrificial activity. If the individual were existentially prior, this would represent either a developmental error or unjustifiable oppression. However, the dynamism which characterizes developing life forms is more accurately explained in terms of individuals being in many ways subject to the needs of transcendent wholes. Altruistic behavior provides evidence of this subservience in the practice of recognizing and assisting others in times of need.

The control over individuals exercised by a gene pool involves no deliberation, being simply a characteristic of any dynamic part-whole relationship. No decision is made regarding who or what shall survive. A gene pool is no more than an adventitious collection of "fittest" genes. The independence of the individual is equally fortuitous. A more accurate interpretation of evolutionary development would be based on the increasing control over the environment exerted by the collective biomass. Interactions at each existential level should be considered as they contribute to the totality that is life. This characterizes the relationship between each aggregation of organic tissue as well as between the various ecologic systems.

The survival and growth of the idea (life) is vested in every instance and at every level, although the obligation is extended in most cases almost exclusively to members of a protected gene pool, which limits the immediate recognition of constituency to related phenotypes. The holarchic model focuses on the conflict between self-oriented desires, for which holarchic existence is responsible, and that of other-oriented feelings which characterize many less sophisticated creatures. The propagation of such feelings is a consequence of the principle of *knowledge*.

Knowledge

Knowledge represents the capacity to experience, interpret, and respond to information. An entity is characterized in one parameter by the extent of its capacity to intercept and experience data transmitted by other existents. Knowledge is not a peculiar characteristic of mind, but is no more than one characteristic—albeit an important one—of each instance of existential being. Knowledgeable entities are so endowed through the influence of several factors; by the physical nature of their manifestation (i.e., their structural elements), by functional characteristics including capacities imprinted by the action of immediately surrounding entities, and most importantly, in the case of living creatures, by information received from their genetic predecessors.

Two classes of knowledge can be identified. *Innate* knowledge is that capacity which is inherent in an entity prior to experience (i.e., knowledge that characterizes the entity in its primitive state). *Learned* knowledge is that capacity which an entity possesses after having been modified by experience and/or maturation. In many situations entities are observed to act without apparent stimulation, as if they were excited from within. Such activity may be said to be based on internally generated information. An example from inorganic matter would be radiation generated by a star. Among organic beings, growth urgency is similarly independent of environmental influence. In each case, internally generated knowledge is involved.[41]

Information refers to the transmittal of knowledge. To inform is to broadcast or express data that reveals the existence, and perhaps some of the characteristics, of an entity. To have an experience (i.e., to be "knowing") is to sense the presence of information.[42] To experience and interpret such information is to be possessed of an altered knowledge state—to have *learned*. The interpretive aspect is involved only in the latter case. Not all entities are capable of interpreting all of the information that they intercept. The human ear, for example, cannot acquire knowledge regarding wavelengths of "sound" outside a limited range of vibrations.

In some cases, such information has a functional characteristic in that it relays a message. The term *communication* distinguishes this aspect of the process when the sender, receiver, or both have a stake in the transaction. Such a data chain characterizes phenotypic-genotypic communion, although the fact that either sender or receiver profits from the experience does not entail the notion that either of the two acts deliberately. At the level of communication, potential for error is introduced. Information may be transmitted and/or received and interpreted by faulty operators, resulting in serious impairment—as when a genetic "mistake" occurs. In this text, communication between the human gene pool and the phenotypes created by the union of genotypic plan and environmental accident are of central concern. The nucleus directs organismic development, and through meiosis (in the gamete>zygote chain), passes knowledge to (communicates with; informs) the genes that specify the characteristics of members of the next generation.

Genes

Genes are agents of a (relatively) perpetual gene pool. They possess the knowledge essential to both self-replication and, most significantly for this text, the creation of proteins that control developmental factors including both behavioral process and the mental elements that direct them. As the unit of information conveyance, the gene has traditionally been thought of as a "segment of DNA that codes for a polypeptide chain or specifies a functional RNA molecule."[43] However, advances in molecular biology have been such that a broader definition is now recommended. Many alternatives have been put forward. Dawkins coined the term *optimon*, which he refers to as "the unit of natural selection,"[44] while Williams refers to an evolutionary gene simply as "hereditary information," ignoring any reference to the entity or entities (i. e., the genes involved) that carry that information.[45]

Genes, of course, possess some form of innate knowledge that makes the reception and interpretation of information possible. Such knowledge, refers to the physical, chemical, and other parameters of all entities. Each gene is, however—as defined above—an agent that simply transmits that specific knowledge of which it has been informed. It is not the purveyor of some type of inherent knowledge and it does not learn from experience. Altered knowledge characterizes only gene pools as a function of the evolutionary process, not as an attribute of the individual gene or gene complex. The knowledge possessed by a gene is, in part, a product of the information on which it is based. However the issue of the source of genetic *knowledge* (not the nature of the *informational process*) must be accounted for, as the function of altruistic behavior is sought.

Genes are assumed to operate on the basis of knowledge that provides the impetus to the maximization of their own fitness. It is on this basis that altruism, or genuinely ethical behavior of any sort, is considered impossible in principle. Such activity, it is argued, would be selected against very rapidly. However, another look must be taken at the knowledge possessed by a gene, and the source of the information that implants that knowledge.

Replicators, interactors, and lineage

Writers of biological texts commonly describe the beginning of the life sequence with a discussion of zygotic genes—the point at which significant transcription is initiated in the embryo's own idiosyncratic genetic equipment. While this is undoubtedly a useful point of departure, it risks creating a misleading attitudinal set. The individual's genetic equipment is well known to be totally dependent on the contributions of its progenitors. Sociobiological literature, in its focus on the "selfishness" of genetic material, pays little attention to the fact that the "self" they describe is a contingent selection of trait determinants that reflects eons of evolutionary development. This phylogenetic flow is channeled through succeeding generations by way of natural and sexual selection processes.[46]

A number of theorists propose that the usual hierarchy of species/gene/organism is no longer adequate to the task of identifying the relationships between elements of life, and even to the describing of those entities accurately. Hull said that "common sense notwithstanding, it [the traditional hierarchy] is 'unnatural'."⁴⁷ Dawkins introduced the term *replicators*, which he, at least initially, assumed could only be genes, and *vehicles* which he referred to as those organisms that transport replicators. "Genes are replicators; organisms and groups of organisms are best not regarded as replicators; they are vehicles in which replicators are carried about."⁴⁸ Hull disagreed, contending that the term *vehicles* assumes that the carriers of genes are no more than passive operators and suggested substituting the term *interactors*. "In my terminology, replicators produce interactors, and the survival of these interactors is causally responsible for the differential perpetuation of replicators."⁴⁹

In developing the notion of replicators, which he had said in 1976 are "anything in the universe of which copies are made,"⁵⁰ Dawkins (1982) made a significant modification in his definition of a gene. "I have previously used the word 'gene' in the same sense that I would now use 'genetic replicator' to refer to a genetic fragment which...does not have rigidly fixed boundaries."⁵¹ Dawkins was suggesting that an arbitrary portion of a chromosome may well be considered a replicator, with the interpretation being dependent on what is being selected for. "If the selection pressure we are discussing is very strong," he said, "the replicator can be quite large and still be usefully regarded as a unit that is naturally selected."⁵² The altering of the label of an entity from the passivity of a term like "gene"—an entity—to the dynamism implied in a term such as "replicator," alters nothing. Hull's use of the term "interactor," in fact, does nothing to solve the problem. The focal point remains a characteristic of the individual.

Lineage is based on the passing on of characteristics of an entity through the reproductive process. *The Random House Dictionary of the English Language* speaks of lineage as a "line of descendants of a particular ancestor."⁵³ Any definition of lineage suffers, of course, from the fact that each specific manifestation of a family line is but a branch of a "tree" that spreads back in time across eons. Thirty generations of ancestors (five to six hundred years for humans—a blip in archeological time) would involve over one billion individuals. To speak of one's ancestry as unique is absurd. The interaction of people is far too intense. Great-great aunts and great-great uncles of contemporaneously unrelated people are undoubtedly often the same individuals, counted many times.

DNA/RNA/protein

A great deal has been learned during the last decades of the twentieth century regarding the possible evolutionary predecessors of current living creatures. The problem of determining the order in which various elements of the reproductive and informational chain came into existence has been vexing biologists ever since. The serious disagreement among biologists regarding the chronology of the elements calls for considerable hesitancy in accepting claims for the priority of any component.

The stance taken by most biologists is that the DNA in each cell is the entity responsible for its own survival through self-duplication, for the survival of the cell in which it resides, and/or that of its copies. The knowledge possessed by DNA is, in fact, expressed in the amino acids which comprise protein.[54] Lawrence argued that many evolutionary questions have been answered by "the identification of internal—that is true—evolutionary processes.... If any of these processes is the 'secret' of life, I would say it's the high degree of self-ordering of amino acids, and the resultant diversity of functions."[55] If such an interpretation is accepted, the principles of order and determinism as they bear on the extent to which evolution can be "free" (i.e., emergent)—that natural selection represents an essentially nonrandom process—must be taken into consideration.[56]

RNA (ribonucleic acid), is believed by many geneticists to have preceded DNA as the repository of genetic knowledge. Eigen et al., for example, proposed that self-replicating RNA came first, with enzymes shortly thereafter, and finally the cells, which hold the materials together.[57] Problems with this view include the difficulty in explaining how RNA, a single stranded template, could be copied at all, since amino acid based protein seems essential to the synthesis. Lawrence pointed out that "A single RNA, even if it could use itself as a template, could not copy its own enzymatic active site and would thus only produce a partial and inactive copy.... No ribosome has yet been found that can catalyze its own replication."[58] It is generally agreed that in living creatures today, RNA has a new function. However, "The essential function of all species of RNA is to mediate the expression of the genetic information [knowledge] in DNA."[59]

In spite of undoubtedly serious differences of opinion, biologists assume that DNA must be a relative latecomer in the scheme of knowledge transfer. The genetic code, which has been carefully analyzed, seems best handled by this aggregation of nucleic acid because of its capacity for self-replication, as well as its comparative stability. As continuing research brings to light more of the activities of cellular elements it is quite conceivable that the starring role so long reserved to DNA may come to be spread more generally among many contributing elements. It may be that more attention should be paid to Fox and others who have argued that the "self-sequencing" characteristic of amino acids makes their part in the creation of the first cells one of far more significance than Darwinists will accept. It may be much easier to recognize the limited function of genetic segments, and to appreciate their subservience to masters past as well as present.[60]

The Role of the Gene in Altruistic Behavior

As elements of the genetic process are considered, it becomes clear that altruistic behavior has its basis in knowledge communicated by a gene pool, just as are other types of deliberate behavior. If such affective experiences as hunger, fear, and emotion in general are products of genetic control, no less can be said of ostensibly altruistic activity. When Dawkins speaks of genes as the "fundamental unit of selection" and of the replicator which has "the extraordinary property of being able to create copies of itself,"[61] he is referencing any gene, or group of genes, that express information in its interaction with amino acids. However, when DNA and/or RNA is analyzed, the conclusion drawn by many biologists is that a "self-ordering " process is being carried on, not by genes, but by the amino acids involved.

Fox and a number of other biologists maintain that the self-ordering procedure of such elements is the basis for the "true evolutionary process," which implies a directional or purposive function not supported by evolutionist biologists. These interpretations provide critical differences in respect to the independence of genetic elements. The notion of the "freedom" of amino acids and the "selection pressure" described by Dawson cannot be easily reconciled. What, then, of the emergence of a moral sense and its expression in altruistic behavior? Does morality suggest purpose? Is it designed to "correct" the "negative" practice of egoistic activity?

Such interpretations have been offered, but they fail on a number of grounds. Most importantly, as was made clear earlier, adaptation is a mindless contingency; what matters is which gene pools are most "fit." The simple fact is that gene pools that include characteristics such as altruism, are at a distinct advantage in the survival race. The lineage of such entities is superior. Not because of the influence of any purpose, but because gene pools whose phenotypes carry a sense of moral propriety are best served. Thus, altruistic behavior must be considered worthy of "moral credit," since decisions regarding the pursuit of ethical behaviors calls for sacrifice that may well be avoided if egoistic interests are given priority.

The budding biologist is apt to be greatly influenced by the expanse of knowledge that has been gleaned from years of research regarding physiological characteristics of living cells. And to the extent that the *biological* aspect is in focus, such knowledge is undoubtedly of value. But this text is not involved in understanding *how* the gene operates, but what *information* it is transporting. Consider several analogies:

> An engineer interested in the structure of a train would focus on the metal, glass, and wood of which it was comprised, the efficiency of the engine, the cooling system, the brakes, and so forth. He would not be concerned with the items being transported, except, perhaps, for their weight, volume and other physical characteristics. *No amount of such information would reveal the contents.* A 6 year old, on being asked where milk comes from, may respond "from a bottle—or a carton."
>
> Such an explanation may be adequate for some purposes, but not for the individual interested in learning the ultimate source of the product. An individual may blame his tardiness for class on the fact that "the car wouldn't start." Perhaps adequate as an explanation to the professor, but practically useless to the mechanic who will work on the machine. Nor is an explanation that "the carburetor doesn't work" much better. When the mechanic learns that "the flutter valve won't operate because it has been corroded" he may have the necessary information. His question may have been answered.

In the case of explanations regarding the role of the gene, even in dictionaries, the simple statement is made that it is "a unit of heredity composed of DNA."[62] (Milk comes from a bottle.) To be more precise, a gene is described as "the shortest length of a chromosome that cannot be broken by recombination."[63] (The car won't start). The information is then refined by the introduction of such terms as "optimons, "replicators," "interactors," and so forth. At the extreme, the concept of DNA, RNA, protein and other descriptors come into play. DNA is "the genetic material of most living organisms.... [It] is a nucleic acid composed of two chains of nucleotodes...."[64] (The brakes are in perfect shape.) How far down that path must one go to learn about the content (information) that is being transported? Even at the end of the sequence, no such information can be discovered! *The wrong questions are being asked!*

The capacity of a DNA molecule to replicate itself, and to express information onto RNA probably represents innate knowledge (as was mentioned earlier). DNA has that capacity at its inception—certainly prior to its function in the genetic process. However, the *data* that inform genetic functions are not inherent characteristics of the protoplasm that they pervade. They represent, rather, an example of knowledge derived from their ancestors; from other members of their gene pool. It is this knowledge, and its ultimate source that is the appropriate focus for the biologist, geneticist, ethologist and molecular biologist.

The ultimate source of genetic knowledge is the gene pool since the fusion of gametic data destroys the individuality of the contributors— much as do the flour and milk that are combined in order to make a cake. If the gene, the DNA, or any component of a living cell were the source of such data, only reciprocal altruism and perhaps kin preference would explain ostensible altruistic activity. However, the randomness of the selection of a desirable mate carries with it the inevitable influence of a "mother-in-law." *Her* relatives (as well as *his*) influence *their* behavior. The zygote is an unwitting, and surely an unwilling, consequence of the association. *However, it is through this relationship that true altruism comes into play!*

Summary

The concepts described in this chapter represent the framework on which an interpretation of altruistic behavior shall be proposed. The major factor involved is that all existents have both part or instance, as well as whole characteristics; that for all existents meaning resides in part/whole relationships; and that in living creatures that relationship is directional, with parts and instances being subservient to the wholes, or systems, that they manifest. Individual phenotypes, in spite of their ostensible independence, are creatures of—and have the function of—enhancing the gene pools that they represent. Activity that is directed toward the self (egoistic behavior) plays a role, in most instances, of enhancing the fitness of a gene pool.

The holarchic principle indicates that at every level of life, be it an individual, a species, a gene pool, a biological phenotype, or a gene complex, some independence from higher existential levels is involved. This freedom from rigid control is essential to the evolutionary process. It makes possible the initiation and development of novel responses to environmental demands, thus providing for "positive" adaptations. However, at the highest levels of life, that latitude leads to an increase in egoistic behavior which represents a threat to the welfare of a group as altruistic behavior decreases.

The discussion regarding the relationship between DNA, RNA, and protein manifests the lack of accord that exists among molecular biologists, geneticists, ethologists, and other life science professionals regarding evolutionary and other priorities of organic components. The scientific method that they attempt to follow is vitiated by disagreement as to the specific function of each genetic component. The "species" is challenged by many as being a specious concept, though some source of the information carried by genes must be sought. In this text, the *gene pool*—an equally amorphous entity—is considered to be endowed with a reality that is manifested in the existence of its instances. The contention that the gene is the unit of selection is, itself, subject to question. Although it is undoubtedly the "carrier of information," that information mandates altruistic as well as selfish behavior.

Chapter 3 Notes

1. Levins & Lewontin (1985), p. 9
2. In fact, evolution may not represent the story of life in the sense that it presumes to. Rosen said: "To me, it is easy to conceive of life, and hence biology, without evolution. But not of evolution without life. Thus, evolution is a corollary of the living, the consequence of specialized somatic activities, and not the other way around.... It may...be more of a property of particular realizations of life rather than of life itself," (1991, p. 255).
3. Dobzhansky (1970), p. 29
4. Fox (1988), pp. 152-153
5. Trigg argued that animals often cooperate with members of other species, indicating that "survival of the species cannot be explained by evolutionary biology," (1983, p. 141). Rosenberg went further, pointing out that "the theory of evolution is...riven with controversy.... There remain members of the biologic community who deny its warrant and reject its claim to cognitive legitimacy," (1985, p. 31). Ho referred to "the inherently fallacious, but unquestioned assumption that...almost all of evolution can be explained by the natural selection of random mutations," (1984, p. 11).

    Such heterodox views are, however, held by only a minority. Most disagreements are reserved to explanations involving various specifics of Darwin's theory. In point of fact, Rosenberg admits that "antievolutionary philosophy has almost completely disappeared within biology. As indeed has almost all philosophy as traditionally conceived," (1985, p. 31).
6. Ho (1984), p. 285
7. *Ibid.*
8. In commenting on the status of the synthetic theory, Ruse said "in recent years the voices of criticism have been raising, (1988, p. 31). Ho and Saunders added that although until recently the neo-Darwinist theory was accepted as ultimate, however, "today...more and more workers are showing signs of dissatisfaction with the synthetic theory," (1984, p. ix). Some have attacked its philosophic foundations, others have devoted themselves to studies in such areas as fossil record gaps and non-Mendelian inheritance. "Still others...have decided to ignore the theory altogether, and to carry on their research without any a priori assumptions about how evolution has occurred," (*Ibid.*).
9. Mayr (1988), p. 527. He insists that the arguments of opponents, "are based on such ignorance of evolutionary biology that it is not worthwhile to provide references to their writings," (*ibid.*, p. 533).
10. Freedman (1979), p. 9

11. Dobzhansky quoted in Kaye (1986), p. 52. Koestler offered a similar explanation. "Before a new mutation has a chance to be submitted to the Darwinian tests of survival in the external environment, it must have passed the tests of internal selection for its physical, chemical and biological fitness," (1967, p. 132).
12. Mayr (1988), pp. 210-211
13. *Ibid.* p. 532
14. *Ibid.* p. 537
15. Day (1979), p. 378
16. Mayr (1988), p. 535. He charges that "rival theories...are so thoroughly refuted that they are no longer seriously discussed.... There is no justification whatsoever for the claim that the Darwinian paradigm has been refuted and has to be replaced by something new," (*Ibid.* p. 12). Those who have attempted to discredit the theory, he says, have only refuted the "reductionist fringe" of the Darwinian camp. "Their failure to appreciate the complexity of the evolutionary synthesis has led them to paint a picture of that period which is at best a caricature," (*Ibid.* p. 536). At another time, he went so far as to say of a prominent biologist who claimed that the only biology is molecular biology, "he simply revealed his ignorance and lack of understanding of biology," (1982, p. 65).
17. Li & Grauer (1991), p. 39
18. Sinha (1984), p. 352. Fox offers an explanation. "Natural selection serves to winnow out the less successful molecular or biological variants, but only after a great deal of internal molecular selection has led the way," (Fox, 1988, p. 153). Furthermore, "large molecules—proteins—*organize* themselves... [They] have the reactive groups that guarantee self-organization. [It] starts at a stage of self-ordering of the monomeric amino acids that go into protein," (*Ibid.* p. 119). At another point he adds, "proteins come...out of nonrandomly interacting molecules.... It is molecules that initiate, 'design,' and 'direct,' evolution," (Ibid.). Fox's work has received considerable applause. Molecular selection is assumed by these writers to be based on certain constraining factors.
19. Rosen (1991), pp. 13-14
20. *Ibid.* p. 14
21. Dawkins (1976), p. 16
22. *Ibid.* p. 17
23. Mayr (1991), p. 116
24. Ridley (1985), p. 3
25. Stebbins (1982), p. 67
26. Ruse (1989), p. 70

27. Mayo (1983), p. 71
28. If one intends to convey by the term "can" (e.g., an elevator *can* transport 20 people) only that it *could*, if the opportunity arose, another meaning is involved; a hypothetical situation is being described. If an adaptation occurs, it does so because the event *must* occur. Its adaptive characteristic is wholly adventitious. Broad agreed, remarking that "natural selection [adaptation] is a purely negative process. It simply tends to eliminate individuals and species which have variations unfavorable to survival.... The plain fact is that natural selection does not account for the origin or for the growth in complexity of anything whatever," (1965, p. 217).
29. Mayr (1991), p. 43
30. *Ibid*. p. 44. Why does he employ the term "necessarily"? And how does the "unpredictability" of evolutionary change rule out purpose? Although no one doubts Mayr's allegiance to Darwinian principles, such a statement can be construed as providing support for at least the possibility of directional evolution; of functional adaptation, which would imply the acceptance of an orthogenetic biological process. Ridley was more specific in his rejection of directional adaptionist notions, contending that "all theories of 'directed variation'...lack an explanation of how the directed variations manage to be adapted to the environment in which they must live," (1985, p. 33). Altruistic, or other-directed behavior can be shown to represent an adaptation that serves the interest of a gene pool. It is a vital aspect of an evolutionarily stable system.
31. Smith (1988), p. 3
32. Bohm (1969a), p. 22
33. Koestler (1967), p. 48. The term was first adopted for use in the text, *Motivation, behavior, and Emotional Health* (Wonderly, 1991).
34. Scott (1986), p. 70
35. Bernal (1967), p. 169. Eigen suggests that "life is a dynamic state of matter organized by information [knowledge]," (1992, p. 15). On the contention that complex systems are self-organizing, and thus not the rare events most biologists assume, Kauffman argues that "life is [the] collectively self-organized property of catalytic polymers," (1993, p. 289). There are, of course, many frankly mechanistic theories; the conviction that man is no more than a complex machine. Zukav, for example, said that "the distinction between organic and inorganic is a conceptual prejudice," (1979, p. 47). Once again, there is wide disagreement. Rosen states resolutely that "the machine metaphor is not just a little bit wrong; it is entirely wrong and must be discarded," (1991, p. 23). The version toward which sociobiologists lean is obvious.

36. Bohm (1969b), p. 41
37. Demes are comprised of small groups of interbreeding individuals (instances) which are those living creatures of which it is composed.
38. Waddington coined this term to emphasize the fact that what is being held constant "is not a single parameter but is a time extended course of change," (1971, p. 12).
39. The notion that a "team" can make a decision—and even worse, concern that some metaphysical entity may be involved—causes philosophers great concern. However, leaderless groups do draw conclusion about a tentative behavior, whether by lot, by consensus, or by acclamation. A president is elected by a *majority* of those who vote—by a group of concerned citizens.
40. Koestler (1978), p. 34
41. Light striking a photoelectric cell, a phototropic plant, or a human eye, causes reactions that are based on knowledge in the recipient. When a knowledge state is activated the response may involve only an awareness of a disturbance caused by some type of stimulus, or it may include some form of action as well. In the case of inorganic matter and mindless life, responses are mandatory. However, among living creatures, to possess knowledge is to be so organized that a response *may* or *may not* occur depending on the circumstances.
42. During the latter part of the twentieth century, it has become appropriate to speak of an *information* explosion. What is usually referenced, however, is the additional *knowledge* that has accumulated. It is this additional, or refined knowledge that is relevant to the role played by the gene. Authors often speak of "information content." Wicken, for example, said that "the program of information required to specify a structure does not translate to the information content of that structure," (1987, p. 17). Eigen added that "evolution as a whole is the steady generation of information—information that is written down in the genes of living organisms," (1992, p. 17). In each instance the term knowledge can be substituted with no loss of meaning. Eigen, (as has become commonplace), was employing the same term to describe the *transmission* of data and the *product* of that process.
43. Li & Grauer (1991), p. 5
44. Dawkins (1982), p. 81
45. Williams (1984). Li & Grauer suggest that a gene be considered "a sequence of genomic DNA or RNA that is essential for a specific function," (1991, p. 5). They contend, that such a function may be carried out without either transcription or translation being involved. This more general

interpretation allows for a wider view of the role played by the central dogma (DNA/RNA>protein) sequence.

46. The influence of genes is a function of their level of sophistication, with clear evidence that they have become more heterogeneous over time. Wills refers to this "wisdom" of the genes, saying that they are getting more resourceful, but "this does not mean that the genes are becoming smarter with time; rather this apparent accumulation of wisdom is due to—and is indeed an inevitable result of—the forces of evolution," (1989, p. 6).

The contention that each gene determines some specific aspect of the physiological—and perhaps psychological—characteristics of individuals has run into considerable criticism from those who contend that genes do not act independently. Williams said, "One must think of the collection of genes within a cell—the genotype—as an integrated unit," (1984, p. 62). Gould added that "there is no gene for such unambiguous bits of morphology as your left kneecap or your fingernail," (1980, p. 91). Kauffman, in agreement with the interpretation proposed here, challenges the entire argument that the development of each individual organism represents no more than the operation of commands issued by various sets of genes. "We have come to think of the genomic system as a kind of biochemical computer which executes a developmental program leading to the unfolding of ontogeny.... "I strongly believe this set of views to be ill-founded, and worse, misleading," (1993, p. 411). The notion that gene complexes issue commands that originate in the genome is, in fact, misleading. The zygote, like each of its genes, is only an agent.

47. Hull (1988), p. 26
48. Dawkins (1982), p. 82. He rejected the notion that individual organisms could be replicators on the obvious grounds that meiosis, in the combining of different genes cannot meet the condition of retention, or inheritance of the specific knowledge carried in the gametes of either parent. (In the case of asexual reproduction, of course, that lineage is preserved essentially intact.)
49. Hull (1988), p. 27. He defined interactors as "those entities that interact as cohesive wholes with their environments in such a way as to make possible replication differential." (*Ibid.*).
50. Dawkins (1982), p. 83
51. *Ibid.* p. 85
52. *Ibid.* p. 89. Although he has been severely criticized for apparently changing his position, it seems more reasonable to assume that such alterations are a product of his continuing study of the issue. In the final analysis, Hull was willing to accept the idea that replication probably functions to

carry out the "bookkeeping" aspect of evolution, "however, in the context of scientific change, omitting reference to interaction leaves out...reference to the entities keeping the books—scientists," (1988, p. 47).

53. *The Random House Dictionary of the English Language* (1987), p. 1117. Hull stated that "lineages are those individuals composed of entities that serially give rise to one another in ancestor-descendant sequences," (Hull, 1988, p. 147). The important point is that individuals, groups or classes cannot be considered lineages as such because they have discrete beginnings and endings. To qualify as lineages they must not only exist at some time in some place, but, said Hull, they must "causally produce one another," (*Ibid*. p. 146).

54. Lawrence stated that "life today depends crucially on proteins...[which] are themselves encoded by nucleic acid, creating a circle from which there seems no escape," (1989, p. 157). Others have taken the stand that proteins, rather than RNA may, in fact, have been the earliest forms of life, because of their capacity to organize themselves. Shapiro contended that: "In the beginning there was protein. Protein begat RNA, and then both begat DNA," (1986, p. 142). "The path from prelife to life, as we know it, passes through protein.... The guiding principle...was the self-ordering of amino acids," (*Ibid*. pp. 155-156). Dyson went a step further, proposing a two stage, or, "double origin" theory, with metabolic capacity preceding replication. "Protein," he said, "is the essential component for metabolism. Nucleic acid [DNA] is the essential component for replication," (1985, p. 6).

He compared protein to "hardware," and nucleic acid to "software." Assuming a two stage birth of life, he said that "the first beginning must have been with proteins, the second beginning with nucleic acids.... The nucleic acid creatures must have been obligatory parasites from the start, preying upon the protein creatures and using the products of protein metabolism to achieve their own replication, (*Ibid*. p. 8). His argument was that "hosts must exist before there can be parasites," (*Ibid*. p. 6).

55. Lawrence (1989), p. 5

56. Fox (1988) contends that although life forms are similar to inanimate matter in having characteristics of order and predictability, the capacity of amino acids, nucleic acids and other molecular forms to participate in preservative and reproductive activities, sets living forms apart. His work is similar in its assumptions to that of Oparin who had proposed decades earlier that life began by the successive accumulation of more and more complicated molecular populations within the droplets of coacervates (stable mixtures that are formed when certain oily liquid is mixed with water).

Oparin had received wide acclaim for his proposal at the time he published his theoretical model. Fox has been similarly applauded for his efforts. However, more recently such models have been vigorously challenged. Rosen claims that many biologists have assumed that if enough properties of inanimate material (e.g., characteristics of form and structure) are combined an organism will emerge. "The idea was that any material system that could be made to embody "enough" of the elements of organisms would then *automatically be* an organism," (1991, p. 19).

Day and others have also criticized efforts to simulate life by constructing morphologically similar entities. "No matter how you look at it" says Day, such models are "scientific nonsense," (1979, p. 318). [Oparin's] coacervates are notoriously unstable, and [Fox's] microspheres exist only in saturated solutions," (*Ibid.* p. 320). Scott was equally adamant. "There is no way for [protenoids] to replicate themselves in a manner that encourages the progressive selection of new and more efficient variants," (1986, p. 380). Like Day, he reviled those who proposed such prebiotic antecedents. "Scientists [who] argue that proteins and...simple nucleic acids have been shown to form and replicate themselves under prebiotic conditions...are simply wrong," (*Ibid.* p. 89). They are repeated as dogma in textbooks "because the authors so dearly want to believe that they are true," (*Ibid.* p. 90).

57. Eigen et al. (1981). Lawrence agrees, stating that: "A popular...scenario for the origin of living cells suggests that replicating molecules (primitive single stranded RNAs)...arising in the primeval soup became enclosed in self-assembling lipid membranes.... A primitive nucleotide-based metabolism provided the energy for replication," (1989, p. 158).

58. Lawrence (1989) p. 159. One suggestion that has been made is that in the most primitive forms of life, *rybozymes*—RNA with enzymatic activity—could have functioned well enough for the needs of primitive organisms, being ultimately replaced by the more efficient proteins. There is some evidence that such rybozymes were, in fact, available.

59. Abeles, Frey, & Jencks (1992), p. 330. DNA (deoxyribonucleic acid), is usually comprised of long polymeric chains twisted into the form of a double helix, which contains the data that informs RNA. The essential function of DNA in cells is to *house* knowledge acquired from previous genomes, and that related to duplication. Wallace said, "*All* genetic information [knowledge] resides in DNA.... The nucleus is the 'home' of most of the DNA.... DNA is information' [while] 'protein is actuality'," (1987, p. 42). Wallace's comment that "protein is actuality" is widely accepted. The process that characterizes the expression of knowledge in most mod-

ern cells involves the transcription of the genetic code from DNA molecules to RNA inside the nucleus.
60. The knowledge that characterizes a gene is apparently spread across its many components. Most importantly, it surely represents the interests of a more comprehensive entity than the phenotype that transports it.
61. Dawkins (1976), p. 16
62. Martin (1990), p. 103
63. *Ibid.*
64. *Ibid.* p. 74

Chapter Four

Altruism and the Motivational Process II

Behavior and the Role of Mind

As it is clear to me that the causal laws governing my behavior have to do with my 'thoughts,' it is natural to infer that the same is true of the analogous behavior of my friends.
Bertrand Russell

In order to provide a model of altruistic behavior that takes into account elements of the evolutionary process, principles of the motivational process and relevant behavior patterns will be described in some detail. Most significant are the factors that contribute to the mental process. These include the role and function of cognitive and affective elements including the nature of beliefs, desire, reason, morality, emotion, and the nature of the deliberative process.

The Function of the Sensory System

The modern synthesis in evolutionist biology assumes that preferential fitness is responsible for the adventitious appearance, and ultimate prevalence, of superior instances of a gene pool. That superiority is manifested in an enhanced ability to cope with a constantly shifting environment. *Every action must be interpreted in terms of its function.* Muscle, skeletal, blood, and immune systems persist in a particular form because of the *function* that they perform in the fitness struggle.[1] They serve activity (including behavior), designed to culminate in some advantageous status. The nervous or sensory system provides the framework on which reason, morality, aesthetics, and other evaluative processes have evolved in higher animals, having reached its peak, to date, in its influence on human behavior. Such evaluative mechanisms must play a role if they are to survive generations of fitness tests.[2]

In the simplest forms of life, what may be described as a "fact-to-act" sequence characterizes all action internal to an organism, as well as that carried on in its interface with its environment. The individual is organized in such a way that internal and/or external stimuli are tied temporally as well as physically to responses. The occurrence of the former demands an immediate response if the organism is capable of responding. There exists no system which makes possible any delay. For such creatures, the perception of a stimulus is sufficient to generate a response.

In primitive organisms no nervous systems, as such, exist. What some organisms have are primitive analogues. The paramecium, for example, has a "neuromotor system" which deals with coordinating the action of the cilia. The euglenia is sensitive to light (phototropistic) and will move toward any light source. In these most primitive organisms both receptors and effectors, if such a label is appropriate, are on or very near the integument. Slightly more complex organisms such as the hydra possess a relatively sophisticated nerve net that is connected to certain muscle cells, and is considered to represent a simple neuromuscular system.[3] In each case, the principle that a functional advantage results in fitness preference is manifested.

The activity of a plant or simple animal that acts automatically in response to specific stimuli in such a way that an evolutionarily stable system comes into being may be totally self-serving. Superior fitness may be based solely on the ability of one individual to overcome others including those of the same species. Mutualistic activity, however, also represents an evolutionarily stable system. Symbiotic activity, such as that of enzyme secreting flagellates that aid in the digestion of cellulose in the intestines of termites, is purely self-serving from the standpoint of each organism. The flagellate cannot live outside the host and the termite would starve without the enzyme. Thus, an evolutionarily stable system is in operation here as well.[4]

What happens that gives reproductive advantage to any of competing forms of life results in preferential selection. Fitness refers, of course, only to that which eventuates. The same process that occurred eons earlier in the evolutionary progression goes on today. Simple life forms, (e.g., single celled organisms) can only be evaluated in terms of their proficiency in the production of descendants, the sensitivity of the organism being still very primitive.

Mind

As organisms have become more complex a more highly sophisticated nervous system whose perceptual centers are located increasingly distant from environmental contact points has developed. The *function* of this interruption is to make a consideration of alternative responses possible. The physical location of the coordinating center has made possible the blending of various sensory inputs. The individual is possessed of a *mind*, which may be defined as "the focal point of a process that involves the cognitive manipulation of perceptual and drive initiated data, and the (affective) convictions so induced."[5] The critical factor is that nothing in the relationship between gene pools and individuals has changed *except the sophistication of the nervous system and the range of behavioral choices available.*

The concepts of egoism or altruism however, are not applicable at the most primitive levels but the nerve tissue can be assumed to have evolved because it performs a function—perhaps serving only the interest of the individual. This freedom from taxic or kinetic rigidity is observed in a variety of invertebrates, in that not all individuals respond in the same manner to the same stimuli. A choice of action is obviously possible.[6]

Some evolutionarily stable systems are comprised of individuals possessed of an urge toward action that will not be in the interest of that organism. Certainly ant colonies exhibit activities that manifest an impulse; perhaps a compulsion, to act in ways that are considered suicidal by human standards. (Wilson has argued that in many ways colonial organisms are most like humans in the performance of activities that benefit the group.) In some cases it appears that the individual must follow a given path. In others, as when a threat to the group arises, there seems some ability to alter a course in the interest of assisting others even to the point of risking personal extermination.

The behavior of these "social insects" (ants, bees, termites, etc.) provides an obvious case of self-sublimation to the interest of a group. Volumes have been written about the various roles played by workers, queens, soldiers, etc. In all instances, the activity of individuals is programmed very precisely, although under certain conditions roles may be altered. Such organisms must be capable of having affective experiences. They are *encouraged* rather than *required* to act in a particular manner. They experience a signal that directs them to take a particular course of action.

While many cold-blooded vertebrates seem to have regressed evolutionarily in their social habits, there are examples of such practices as protecting the young even among reptiles, with cobras, for instance, being known to defend their brood against potential invaders.[7] This represents an early sign of the impact of intelligence and the reasoning process, as well as an urge toward behavior that includes a risk factor.

At the highest levels of social development the conditions necessary for the occurrence of genuine egoistic and altruistic behavior emerge. Since the notion of self and the motivational process are ordinarily associated with an awareness of one's existence, it is necessary to consider the processing of data at the mental level. Such concepts as *desires*, which are the mental representation of drives, and *perception*, which includes the interpretation of sensations, come into play. Intellectual skill has also become a significant factor. Although intelligence, which may be defined as "the docility of a learner to new information,"[8] is a characteristic of all existents, in the living it serves a critical purpose.

The emergence of mind and its components has resulted in the appearance of many species of life that are simultaneously capable of the most rapid and efficient self-enhancement and the proclivity for self-destruction. Humans are, until now, the ultimate example. The tendency for parts of the living chain to possess varying degrees of autonomy is clearly manifested. At the level of mind, action occurs on the basis of deliberation. The desire for power and safety interfaces with the desire for meaning and purpose. Choices pit individual gain against profit to others. Demand for immediate gratification is set against considerations of health and safety. The challenge is to find a solution to this inevitable anomaly; to seek an interpretation of those behaviors that appear to represent the sacrifice—and/or risk—of one individual's interests in the pursuit of those of others.

Deliberations are based on the delay potential that resides in mental processing, and the interplay of emotions which are influenced by such factors as attitude, experience, deprivation level, and perceptual set, as well as on one's conviction regarding the capacity and opportunity to act.[9] The capacity to perform this delay, and to more efficiently weigh the quality of those factors that are relevant to the deliberative process, resides in the performance of a cognitive/affective system.

Cognition/affect

Cognition embraces all of the *actions* of mind including, in addition to conceptualization, such processes as reasoning, abstraction, generalization, recall, and memorization (but not abstractions, etc. as such). It also includes the analytic and synthetic functions. Beyond this, the cognitive process deals with the interrelating of information from all data inputs (i.e., comparing percepts with each other and to desire, as well as to the various judgmental scales). However, it does not represent the experience described as *awareness*, nor does it represent the possession of knowledge but only the method by which information is processed. It refers rather to the manipulation of the material *about* which one is aware.[10]

With the evolution of the cognitive process, increasingly finer distinctions can be made between concepts. The evolution of that function has resulted in the erroneous conviction that behaviors based on thinking (cognizing) are an evolutionary improvement over those driven by emotion. However, what the cognitive process represents is a technique for data manipulation that provides an advance over instinctive knowledge. It is not in competition with, or a substitute for, emotion, but rather represents a contribution to the emotional process. In lower organisms, the individual has innate knowledge of what to do, which fixes responses within a narrow band. In higher forms, learning occurs partly due to the cognitive process. Cognition relates to knowledge as the architect does to architecture, being only an instrument of the process; or as listening relates to hearing, the latter being essential to a knowledge of what is being heard.

Affects are feeling states that result from the activity of internal and external information vehicles. They are peculiar to, and, in fact, define the nature of consciousness. They function to quantify and qualify contemporaneous as well as imagined or recalled situations (which challenges the absurd claims for unconscious desires, emotions, etc.) Feelings of desire, attitudinal and sentimental reactions, temperamental and mood states, emotional experience, and the awareness of the perceptual senses all impinge on the individual as forms of belief or conviction. A distinction must be drawn between a cognitive action that produces a concept, and the appreciation, understanding, or awareness of that concept, which is a form of affect.

To perform the inductive operation is a cognitive act, but the conviction that the sun will rise tomorrow because it always has represents an experience that transcends the intellectual exercise. The ability to perform a cognitive act is one form of knowledge. It represents, however, knowledge only regarding its function, not of the quality or even the quiddity, of its products. "Treeness" and "sweetness" are generalizations from experiences as identified by separate senses. To say that a tree is a form of plant life is to move from one conceptual level to another which is a cognitive process. However, to conclude that the concept is appropriate in any particular situation is an affective experience.

Belief

When knowledge is activated as a mental phenomenon, knowing occurs in the form of *belief*, which is based on innate knowledge, or cognitively processed information. It is a *class*, or *type*, of knowledge, which is the epistemological limit of mental experience, regardless of whether the data are drawn from internal or external sources. To believe is to have a feeling about the legitimacy of contentions regarding processed information. This form of knowledge is unique to those forms of life possessed of an affective system. Although it is based on the organizational state of the individual as tempered by learning and experience, as well as by some forms of instinct, it is held as a matter of degree.

Beliefs are influenced by idiosyncratic characteristics of each believer, and are classified in terms of level of strength, or degree of assurance. *No instance of mental knowledge transcends this limitation.*[11] Stroud concluded that "no adequate theory of the nature of belief has been given to this day, and that is probably because it has been investigated in virtually complete independence from the notions of passion, desire, will and action."[11] His criticism is valid. Every philosopher who has attempted the task has become mired in the conviction that belief and knowledge are distinct concepts.[12]

To claim knowledge of the process of deduction is to profess an ability to perform an operation (to know how). But, a problem would arise if one claimed that the application of the deductive process mandated acceptance of its conclusions without the mediation of some feeling state. Nevertheless, belief with all of its limitations, represents the most sophisticated expression of knowledge, being a flexible and resilient medium for the interpretation of information. The deductive argument is convincing because it creates a positive affective situation. It seems correct; it makes sense; it persuades. Inductive "leaps" are similarly convincing.

Desire

At the level of mind, "drive" knowledge is expressed as information that impacts the affective system in the same way as do percepts.[13] Instead of being driven to a response in the presence of an urge, the individual is aroused by signals of varying degrees of force; one class of which is desire, that encourage the recipient to do something, to take action; to change position, to go somewhere, although they specify no particular procedure. Signals take the form of recommendations rather than commands, and intensity is a factor in the deliberative process. Since desires function to encourage activities that enhance survival and growth, and since the survival of life is commonly considered to be desirable, they represent a positive force. They are, however, positive only in their function; not as they are experienced. The awareness of any type of urge to action is a signal to rid oneself of that feeling. *All desires are experienced as negative affects;* irritating or annoying experiences.

The term "desire' is commonly associated with the notion of "want," which means to lack. To say that individuals desire something means that they lack it. They are receiving a signal that tells them so. Consider a rheostat that provides information to a heat unit. When a preset point on a scale is reached, a signal sounds. That signal is a form of negative feedback. It is designed to cause a response. The individual is similarly prompted by signals that are peculiar to various types of deficiency.

At this level the Freudian interpretation is applicable, in that homeostasis refers to the dissipation of uncomfortable feelings. Stimulation seeking, the widely observed phenomenon which has been used to challenge the psychoanalytic principle, is no different than that of food seeking. Lack of stimulation causes a feeling of distress differing only in quality from a feeling of hunger. In discussing the orthodox conception of drives (referred to here as desires) White agreed, saying that "the tension of an aroused drive is interpreted as unpleasant, at least in the sense that [one] acts in such a way as to lower the drive."[14]

Classes of desire. A significant factor in the attempt to understand altruistic behavior is the relative precedence of several types of desire; the order in which they tend to be addressed, and the comparative potency of their signal. Most powerful are those urges whose gratification is related to aggression or acquisition. They include such desires as hunger, thirst, and sex, which arise spontaneously as drives in both primitive and sophisticated forms of life and are based on chemical tissue change which occurs as a function of time. They may be described as *self-assertive* or egoistic. The goal of related behavior is the enjoyment of an experience.

A second class of desire includes those related to threat. Such desires may be defined as *self-protective*. They are aroused only when the integrity or safety of an individual is at risk. They are, thus, classified as contingent. Confused perceptual states, distortion of images, and any inability to interpret information causes an alarm system to function which is experienced as an urge to alter the situation. The signal is one of pain or discomfort, and relief rather than pleasure is sought. In the case of escape behavior, the need is not for the behavior but the outcome.

On the basis of the reasoning involved in the determination of the role of desire, and its ubiquity as a behavioral element in the case of assertive and protective urges, behavior directed toward the welfare of others must be interpreted as based on the awareness of the same type of negative signal (a desire). This third class of desire can, thus, be identified as *self-transcendent*. The goal of activity based on that desire is related to the enhancement of the gene pool that is the originator of the message. And that message is the same for every organism possessed of an affective system. The behavior associated with its resolution is no different in principle than that of egoistic urges.

To appreciate the fact that desires are common to all forms of sentient life, it is necessary to look beyond the term; to focus on the experience that one undergoes. It is extremely difficult to communicate the precise nature of that experience. What sort of feelings do people have when they are hungry or thirsty, frightened, or concerned for a loved one? And why should it be presumed that such feelings are uniquely human if their purpose is to encourage behavior?

As bizarre as it may at first seem, the experience that a housefly undergoes when it observes a bit of sugar after having been deprived of food for a period of time, is no different *in principle* than that of any other organism, including the human. A genetically implanted (negative) affective signal encourages specific types of behavior in appropriate situations. When a chicken becomes aware of a hitherto unseen hawk it undergoes some affective experience. It desires something. In this case, to get to a safe place. When a hungry person sees a desirable food an alteration in that person's biological organization is experienced. The individual desires that food.

Such states are functional. The bird is encouraged to fly north or south, the snake to recoil, the antelope to follow a path carved out by generations of ancestors. The principle can be extended to explain the action of the simplest organisms. Each behavior occurs as a result of the experiencing of a negative signal, and in many cases in the anticipation of a positive feeling. Most importantly, individuals need have no awareness of the consequence of their actions. It is sufficient that they respond to signals, and attempt to take ameliorative action.

The focus of this text is on the self-transcendent desire, and the emotion aroused when one becomes aware of a situation involving the needs of other people. That awareness, when accompanied by the belief that one has the capacity and opportunity to take action, is often sufficient to trigger altruistic behavior; to encourage action that is not based on assertive or protective urges. As is the case with all desires, in many situations only the behavior is in focus. The individual may not—indeed need not—attend to the consequences. Altruistic behavior may call for self-sacrificial activity. However, from the standpoint of a gene pool, the pain associated with behavior that results from efforts to assist others is of relatively little consequence since it is through such suffering of individuals that the fitness of the gene pool is enhanced.

The critical distinction between transcendent and other desires lies in the fact that the former is aroused in situations which portend the occurrence of outcomes resulting from the experience of giving; sharing; sacrificing; whether such outcomes are known by the behaver or not. A need for an individual may be the performance or the outcome of a generous act, though such action is ordinarily a cost involved in achieving a beneficent outcome. However a desire for *something* is the source of altruistic as well as egoistic behavior.

In the case of human altruism, the stimulus is experienced in precisely the same way as it is in lower animals. One has a feeling that self-sacrificial behavior is called for and is willing to accept the risk, or loss, involved. If the same principle that explains the risk taking behavior of the killdeer when its young are threatened is to be applied to more sophisticated forms of life—including humans—it must be concluded that there exists both a desire for the welfare of others that is sufficiently powerful to encourage the willingness to risk, and, at some level of evolutionary development a sense that evaluates such behavior as morally appropriate. (The concept of *willing*, will be developed as an aspect of the behavioral process.) The contention that such influences are involved is based on the assumption that humans function in much the same way as do less sophisticated creatures. While a preference for serving the needs of relatives may be involved, at simpler levels of life a motivational process based on reciprocal altruism seems highly unlikely.

The sociobiologist's contention that risk taking behavior is a form of reciprocal altruism; that the individual must anticipate some recompense may be true for mature humans (and undoubtedly often is). However, it surely cannot be applied to the behavior of infants and simpler animals. It would be presumptuous to contend that the baboon sending a warning call takes into account the possibility that other animals will reciprocate. Such an intellectual exercise is probably far beyond the capacity of such animals. The urge to act in ways that enhance a gene pool on the basis of self-sacrificial behavior is generated out of a fitness formula that is evolutionarily stable for that gene pool. The individual is only a component in a dynamic system; an agent in the process.

Desire as instinct

It is common practice to explain unlearned behavior as being based on instinct. However the use of that term to explain the behavior of animals—and even some of those of humans—is based on an extremely narrow interpretation of the concept. Instincts are, in fact, of two types. Instinctive *behavior* refers to an innate ability to perform a certain action, such as the building of a nest or web. Instinctive *knowledge*, refers to the capacity to appreciate the significance of a percept; to interpret information that appears in the form of a signal; to recognize the conditions under which a behavior is to be performed.

Desires represent one type of instinctive knowledge. Although individuals learn techniques for gratifying desires, and even conditions under which particular activities are appropriate, they do not learn to feel hungry, thirsty, or angry. Rather, they learn the kind of situations and events which reduce discomfort, or are experienced as enjoyable. This information/knowledge/behavior/outcome chain characterizes all sentient creatures, differing only in the degree of freedom from fixed responses that evolutionary development has engendered. The behavior of such creatures provides evidence of the influence of a desire; of the experiencing of a negative signal. When one is attracted by the smell of food, it is obvious that the purpose of such an experience is to encourage eating, *to assuage the feeling of hunger.*

Needs

Although desires lie at the root of behavioral impetus, they ordinarily become identified with particular forms of gratification. In contrast with desires, individuals are attracted to *needs* which represent positive experiences. The relationship between desires and needs may be visualized as a continuum, with desires representing a negative valence, and needs a positive force. To satisfy the desire to eat, some kind of food is *needed*. The sex urge calls for some form of *needed* action. At the simplest levels needs such as the appropriate sex object or potential food are instinctively known. The analysis here will, however, be restricted to those experiences which individuals learn to pursue because of the belief that they have desire satisfying properties. Such entities, experiences, and events shall be identified as *learned* needs and bear the same relationship to desires as *primitive* needs do to species survival and growth.

Needs are considered in a value hierarchy, with alternative choices being continuously compared. In a motivational sequence needs under consideration rarely represent the most desirable, but rather the best available, and/or least costly alternative. Thus, a teenager typically pursues any type of food that may contain sugar. An example of the sequence as it relates to food and hunger follows:

Primitive needs - comestibles are needed to sustain life
Instinctive needs - comestibles innately known to satisfy hunger
Learned needs - comestibles that one learns will satisfy hunger

Since needs are not only for objects, but also for experiences, events, and behaviors, whenever the term is employed the appropriate referent must be taken into account. The critical factor in each behavioral episode is that desires are *always* negative experiences, while needs are *always* anticipated to be positive.[15]

Evaluative procedures

In deliberating about which of alternative behaviors is to be carried out, one evaluative technique is that of *reason*, which refers to, "the appraisal of probabilities where behavior is involved."[16] These "probabilities," refer to the level of confidence one may have that a behavior will have a propitious outcome. To reason, as a cognitive process, is a form of behavior. However to contemplate the reasonableness of a behavior is an affective experience. Hume contended that "reason is the discovery of truth or falsehood."[17] The reasoning process, however, does not discover truth, it *infers* it. Most importantly, no collection of "facts" is sufficient to determine whether a behavior is or is not reasonable. All such determinations are based solely on the individual's estimate of the relative strength of desires and the extent of the cost associated with behaving.

Many social scientists have taken the position that reason is the highest mental function. Koestler held that the rational capacity in humans represents a possible release from the control that emotions exercise over behavior in more primitive animals.[18] The thrust of such arguments is that since civilized people have the ability to act more rationally, they can be expected to exhibit greater moral solicitude. However, such thoughtfulness is a function of a second evaluative scale which "communicates...the propriety of situations and events as to their self and general life enhancing characteristics."[19] This (moral) sense represents an entirely separate and distinct assessment scale. Unfortunately, improved reasoning potential conflicts with altruistic considerations in many situations.[20] Albert and Denise summarized the view of those who contend that reason and morality are linked, saying:

> When we endeavor to fill in the blank places in our moral theory, to eliminate as far as possible contradictory directives for behavior...and when we endeavor to know why we think an ideal or moral judgment is correct, we have made a good beginning in the direction of applying reason to the moral life.[21]

Socrates argued that since no one voluntarily acts in an evil manner, a person governed by reason can avoid depravity. But to be governed by reason may lead one in a totally opposite direction. It cannot be shown, for example, that to take advantage of an illegal opportunity to profit, where there is virtually no possibility of being apprehended, would necessarily be unreasonable, though it may certainly be considered immoral. Such interpretations represent one of the serious flaws involved in the assumption that naked reason should lead to the performance of altruistic behavior. They represent a common example of the confusion between the concepts of reason and morality. A behavior that makes possible the gratification of a desire may be seen as reasonable to the extent that it can be carried out at an "acceptable" cost—except that it is patently immoral. Here the conflict is obvious. However, even in a situation in which the meeting of a desire is both consistent with an ethical code and appears to be reasonable, separate characteristics of the act are being considered.

Neither morality nor reason is a class of desire or a type of need. One does not behave in order to be "good" or "reasonable." Behavior is driven by the strength of desires, the need value of relevant behaviors, situations, or events, and the anticipated costs involved. Reason and morality—like desires—are a class of affects. They are appropriately described as *feelings* or *sentiments*.

There is a tendency to assume that some behavior is driven by the rational or moral sentiment; that one acts in a particular situation because they believe they *should* do so. However, although both desire and evaluation are elements of a behavioral sequence, the distinction between wanting to assist, and feeling that one should, are separate features. Altruistic behavior is influenced by both factors. This is not, of course, to argue that a gene for altruism emerged.[23] What has evolved is the moral sense, which like that of reason—is a guide in the selection of behaviors.

Emotion

Emotions serve the *function* of tying innate urges and evaluative processes to perceived, recalled, actual, or imagined, environmental conditions. Just as belief is a subset of knowledge, so emotions are one of many types of affect. They are a form of knowledge which is activated by an awareness of the interaction between desires, associated needs, and the conceptualized environment,[24] and are experienced *only* when desires and elements relevant to their satisfaction are involved.

Emotions are influenced by the evaluative scales and, during the deliberative process, by belief about capacity and opportunity to behave. They are experienced both positively as behavioral prospects are encouraging and negatively as desired behaviors seem improbable, or impossible. Emotions, like desires, are experienced only at the time that a behavior or a situation is under consideration. One cannot remember an emotion; only the factors that caused it. The recall of an event may cause an emotion, but that experience is related only to the present.

Although deliberate action may be taken in order to achieve an emotional state, in emotionally healthy individuals negative emotions are never sought as ends in themselves. They are either aroused by unavoidable situations (where they are designed to prepare for corrective action) or are tolerated as part of the cost associated with some competing present or anticipated future positive feeling. The contention that fear, for example, is apparently sought in high risk entertainment can be shown to be erroneous by considering whether dangerous behavior would be indulged if the stimulative aspect were removed. If no concomitant exhilaration is involved, personality disturbance may be confidently predicated.

In the evolution of species, the bonding of an emotion and an experience represents an example of genetic learning. Since certain interactions between phenotype and environment result in outcomes that have preservative and growth aspects, some of the characteristics of the environmental contribution are invested with affective qualities. However, if the only occasion on which an individual could experience an emotion came after an activity, it would be of little value to the gene pool. Although the consummatory emotion (the emotion that accompanies a behavior) would appear to represent positive or negative "reinforcement," it could not lead to a repetition or avoidance of the act. It is essential that the individual be capable of anticipating the reinforcing state and of experiencing an anticipatory emotion (the emotion experienced when contemplating a behavior) that encourages action.

If the feeling of guilt that attends an action that one believes to be immoral, (such as the gratuitous mistreatment of another person), caused no change in future action, it would represent punishment of the type described in the Old Testament of the Christian Bible. It would function as capital punishment does when the notion of retribution is involved. The call for the payment of "An eye for an eye" assumes that the treatment of the guilty need not be related to a change in behavior but is just in its own right. The significance of both anticipatory and consummatory emotions as they relate to a behavioral sequence must be taken into account.

Motivation

The motivational process flows continually from an emotional state through deliberate activity, which is directed toward the optimization of an affective state. It is characterized by the initiation and culmination of behavioral sequences, involving all of the factors that influence a potential behavior. This definition distinguishes motivation from desire, drive, and any of the judgmental scales that are sometimes inappropriately considered to be the independent cause of a behavior. The individual is capable of considering many options, both as to the urges on which to focus, and which behaviors to employ in attempting to assuage them. The order in which elements are considered varies both in terms of the class of desire and the perceived situation.

In the case of the deficit desires, (e.g., hunger, thirst, and sex), the strength of the urge is a function of time and the individual becomes increasingly agitated by a desire for gratification. In many cases, the awareness of a *need* will focus attention on a desire and action to provide satiation will be encouraged. However, *needs do not cause or create desires.*[25] Protective and transcendent desires remain quiescent until and unless they are provoked. Here, the conceptualized situation or state of affairs arouses a desire, followed by attention being focused on potential needs. As with needs, situations are only the vehicles for the arousal of desire.

If a motivational sequence is to be initiated, many contributory components must be involved. However, the arousal of an emotion, the wedding of conceptualized desire, need, and environmental factors, is only one step in the process. Emotions are necessary but not sufficient to trigger the motivational sequence. None of the many elements of a motivational sequence *alone*, provides the data essential to fully determine a deliberate act. Sex and hunger represent only the urgency of desires of a physiological nature, while integrity demands deal with the desire for stability. In all instances, belief regarding one's capacity and opportunity to act are also involved. The focus of this text will be on the self-transcendent desires and the emotion aroused when one becomes aware of a situation involving the well-being of others. That awareness, when accompanied by the belief that one has the capacity and opportunity to take action, is often sufficient to trigger altruistic behavior; to encourage action that is not based on assertive or protective urges. Affects involved include a willingness to risk or sacrifice one's interests to the benefit of others, and a desire for such action to have a beneficial outcome.

The self

The self in the conscious individual represents both the cause and the result of the interplay of genetically fixed dynamisms and environmental opportunity. As a cause it is the source of, and responsibility for behavior. As a result it is a product; a contingency. The latter interpretation is apt to cause concern among the advocates of a "free will" philosophy who are convinced that the self only directs, rather than responds to the impetus of physical and mental prodding. At this point, however, the only interpretation intended is that an entity (the gene pool) be seen as the more appropriate source of behavior; that the self is no more than a label applied to an agent.

At the most sophisticated level of phenotypal existence, the sensations of *pain* and *pleasure* are experienced by a self. As intellectual capacity increases, more techniques for finding and sustaining pleasure, and for avoiding pain, are learned. Such intellectual development in many instances results in a diminished interest in the needs of others. Not only is sensation localized, but the partness aspect of individual existence is not so clearly manifested. Individuals can be perceived acutely, but the generality of the form of life represented is apt to be conceptually blurred. People capable of experiencing positive and negative affects are equipped to meet personal survival and growth needs more effectively than those forms of life not possessing such sensitivity. Since self-directed feelings are more intense than those related to the interests of others, the amount of altruistic behavior is often reduced. The most negative aspect of holarchic existence is manifested.

The necessity of assuming a relationship between individuals and higher echelons of life is, of course, not universally accepted. The notion of an individual soul for each human being and the many philosophical arguments on the value of individual life imply that something exists within each living being which is independent of any more comprehensive existent. Portman, in his concept of "interiority" suggested that living organisms "overcome their corruptibility through reproduction and hereditary transmission...[and] are capable of relatively autonomous activity."[26] Such interpretations exemplify the tendency to point toward the biological individual as the proper focus of selfhood. The organism is said to overcome extinction by reproducing itself. This view represents an example of the inaccurate suggestion that the individual is more than an agent of a process with which it is invested by its genotypic forebears. It does not reproduce itself, but is only one of the many elements involved in the reproductive process.

Identification

The dimensions of a self are extremely difficult to identify, in part because of the flexibility of the experience. Clearly, the self is not diminished because a part of the body is lost or nonfunctional. The self of a one-armed person is quite complete. When one speaks of a self that includes more than the physical body, however, the concept is more difficult to accept. If John is known as a "family man," or, in political parlance, that Arthur is a "party man," the reference is to an individual who gives priorities to some person or organization other than the biological self, and this is an appropriate interpretation. The man who gives his life for his child can be said to recognize a self which transcends his physical being. Although the locus may be biologically determinate, the extension is not limited to the physical body.

While it is proper to say that the individual who gives, or shares, is identifying with a larger system, it is equally meaningful to say that the self includes more than the physical person. The self expands as the locus of behavior shifts from needs based on the self-assertive and self-protective urges to those associated with the self-transcendent desire. Although awareness of the self is located within the biological entity, the object of such awareness can range from the body itself to the totality of life. "Who am I?" becomes: "What is the self of which I am aware?" The answer defines the self as *that with which one identifies*. This incorporation of others into the self as an act of identification represents a recognition of gene pool superordinacy rather than a creative enlargement of the self. For this reason, the "expanding self" described by the existentialist, which never moves its existential center from the physical or biological being, cannot be accepted.

There is considerable opinion to the effect that—at least among humans—a well developed identity is essential to effective interpersonal effectiveness. Guisinger and Blatt argue that "an increasingly differentiated, integrated, and mature sense of self is contingent on establishing satisfying interpersonal relationships; conversely the development of mature relationships is contingent on the development of mature self-identity."[27] Interpretations of this type provide evidence for the existence of both altruistic and egoistic tendencies, unless the sociobiological contention that all interpersonal relationships are developed in the interest of using such contacts for personal gain is accepted.

The relationship between *selfhood* and *selfishness* is very complicated, since the individual who acts in the interest of those at whatever level a sense of identification is experienced may technically be said to be behaving *selfishly*. Altruism may be described as no more than selfishness directed toward an extended self. But that is to misunderstand the distinction. To be aware that one possesses a self is morally neutral. However, to act selfishly is to behave in a way that is considered immoral by any definition. Just as killing, for example, is a neutral term, (animals are killed to provide food; "enemies" are killed to protect one's family), murder implies the violation of an ethical code. Similarly, selfishness is a term that is appropriate to behavior directed toward the biological self or, in some instances to those closely related, at cost to others of no obvious relationship.

Selfhood and identification are closely associated, and that knowledge of the relationship is, in part, a form of moral awareness. The scale of morality and the desire for identification proceed from a level at which parts and wholes are indiscriminable. In the simplest life forms, no behavior is possible, and total identification obtains. In humans, the existence of much competing data make identification a sometimes serious problem. With "whom" or with "what" the individual identifies becomes a significant issue.

Responsibility

The term responsibility is defined in one sense as "chargeable as being the author, cause or occasion of something."[28] Thus, on many occasions responsibility may be ascribed to the performance of an inanimate object. It identifies a critical element in the causal chain. In its primitive form it conveys no moral reference. If a car is involved in an accident, it may be discovered that the brakes "failed." It is appropriate to state that an ineffective brake system was responsible for the mishap, assuming that no deliberate damage was involved. A vital characteristic of the holarchic chain is that:

> *Responsibility devolves onto the whole aspect of the last holon (entity, situation, or event) in a causal sequence.*

The value of such an assignment of responsibility is, of course, somewhat vitiated by the fact that either the ontological existence of a "last holon" may be assumed, or the epistemological limit of a sequence may be accepted. Consider Franklin's maxim:

> For want of a nail, a shoe was lost; for want of a shoe a horse was lost; and for the want of a horse the rider was lost. [For want of a rider a battle was lost; for want of a victory a war was lost; for want of a triumph a kingdom was lost.] [29]

If the causal sequence is followed to its ontological limit it is clear that the nail *in its whole aspect* is responsible. If, however, only the groom involved is known, he may be considered the responsible agent. The epistemological limit carries the onus. If someone had deliberately destroyed the nail another link would be added to the succession. In analyzing the steps involved in arriving at a decision, the role of the self *as a whole* in the deliberative process must be considered. Thus, while the parts of a self may be involved ("the devil made me do it"), they cannot be assumed to be responsible. The focus must be on the whole aspect of the last member of the holon chain. The integrity of the whole must be maintained.

Deliberation and decisions

Decisions are based on the interplay of emotions which are influenced by such factors as attitude, experience, deprivation level, and perceptual set. This process produces a constantly fluctuating decision state which is the climax of the deliberative process. It compels behavior at some appropriate time. This follows from the fact that a decision is not free of the impact of deliberation, just as one is not free from "hurting" after being struck, or of believing (or not) that something is the case after having thought through an issue. All decisions result from a comparison of the anticipated cost that is involved with the potential satisfaction value of the contemplated behavior. Whether the choice is between positive options (attending the theater), negative options (cleaning a basement), the situation is the same. The behavior that occurs is fixed by the data.

The decision making process is a uniquely conscious mental function, being performed in response to commands issued by the self. Its peculiarity in its relationship to behavior is that the elements of desire and evaluative processes are included as determinants. The determinism that is involved refers, however, only to the moment of decision, and to the focus of the decision, its fixed aspect being in relation to the totality of the activity involved. In the case of the deliberative or decision making process one is free of the constraint of an automatic or taxic response.

The act of making judgments about the value to be assigned to needs and costs and to determine the strength of a desire is the responsibility of the behaver. (The self is the whole aspect of the last holon in the chain.) Rational, moral, aesthetic, and other evaluative mechanisms provide for a broad range of responses. However, although the deliberator takes such affective influences into account, the decision is as fixed as is the action of a flower as it bends toward a light source. The fact that desire is involved does not make the decision free, but adds a determining factor. The freedom that is involved is based on the fact that the *behaver* assigns values to need and cost factors. Thus, deliberation represents a holarchic process to some extent free of the influence of the gene pool that is the source of its genetic content. It provides evidence of the generation of freedom in a deterministic universe.

Behavior

Behavior occurs on the basis of an affective state resulting from the weighing of relevant factors, (desires, needs, costs, etc.), the belief that one has the capacity and opportunity to act, and the conviction that such action will either be salutary in itself, or lead to desired consequences. Behavior, like all genetically driven activity has a *status* function, which is to achieve an optimal outcome for the gene pool that the individual manifests. In the pursuit of this goal, its immediate function is, in many instances, to meet survival and growth needs of individuals; to focus on the self-assertive and self-protective desires.

Some behaviors are based on sheer desire (e.g., hunger), while others are driven by the enticement of a specific need, (e.g., a particular food). The term *want* is employed to embrace both concepts. *Costs* refer to the negative factors that are associated with a potential behavior; all of the needs and desires sacrificed if a particular behavior is consummated. Behavioral considerations are not made in isolation. Deliberation *always* involves alternatives. Choices must be made in each instance. Needs are projected against costs, all of which are psychologically determined. A cost, such as for example the price of an item, is totally subjective. What one can "afford" is purely idiosyncratic; a determination made on the basis of belief about one's resources.

In determining whether to pursue a particular behavior, both the level of cost, and the value of needs are ordinarily anticipated. One *believes* that the anticipated price will be manageable. One *presumes* that a particular need is apt to gratify the desire under consideration. Only when the individual is in the act of behaving are the value of such factors known to any degree of certainty. Although needs refer to entities, behaviors, situations and events that assuage desire, the *goal* of behavior is not the experiencing of a need as such but the cessation of the influence of a negative signal. The aim, from the genetic standpoint, is the completion of an act rather than its performance; of the return to a relatively quiescent state. The source of such a process is not to be discovered in the gene, but in the gene pool which is the progenitor of all behavior. The function of deliberate, as well as non-deliberate activity is geared to gene pool enhancement.

The Behavior Adjustment Paradigm (BAP) shown in Figure 4.1 is adapted from the text *Motivation, Behavior, and Emotional Health*.[30] It represents a general model which is applicable to all behavioral modes—behavior incidents, total behavior sequences or instrumental or intrinsic behaviors, and/or behavior patterns. On the model, letters designate scale points, while numbers denote the behavior adjustment location. The diagram refers to the focused behavior or outcome. The location of a desire or cost on the model is purely arbitrary in that there is no "proper" or "correct" location in spite of a commonality of opinion. What is considered serious by one individual may be of little consequence to another as judgments are based on experience and other factors.

The behavior threshold refers to the point at which netwant (the sum of all wants) and netcost (the sum of all costs) are of equal strength. The upper left area (OAF) represents the decision to perform a specific act, while the lower right (OCF) represents the decision to reject that particular action. The areas OAF and OCF represent the intensity of the emotion accompanying the relevant action. The area bedF identifies the location of behaviors or alternatives which are associated with psychological discomfort. The dividing line is obviously not distinct. The increasing width of the crosshatching shows that the discomfort is more intense. Emotions approach this level as costs and needs become mutually higher. If either need or cost value is very low, no distress is involved.

Behaviors may be defined as *intrinsic*; where the action is valued in itself (e.g., playing golf, making love, or painting a picture) and *instrumental*; action carried out in order to facilitate the occurrence of another action or outcome. Working at an unpleasant task or sacrificing for another individual represent examples. The term instrumental *behavior* is employed in spite of the fact that it represents a cost in a total sequence, in order to distinguish it from unmotivated activity, such as a reflex, or a taxic response. Since all behaviors have a status goal, activity that is intrinsic to the behaver, (satisfying in itself), is instrumental to some existential situation or status. The reason that intrinsic behaviors are enjoyed is because of their relationship with some outcome. *Behaviors are never ends in themselves.* Their function is to optimize the probability of the occurrence of desirable *status* conditions, whether recognized or not.

Figure 4.1
Behavior Adjustment Paradigm

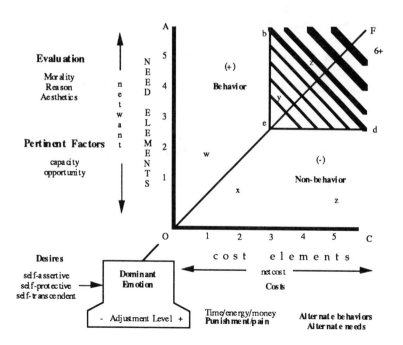

In the case of instrumental behaviors, there is a temptation to assume that the model employed here is inapplicable, as when one deliberately undergoes a surgical procedure. However, the goal of the behavior under consideration is a condition of improved health. The appropriate analysis involves a consideration of the total behavioral sequence. The surgical experience is a *cost* associated with the process. If the goal were the surgery it would certainly be avoided. This misinterpretation will be seen to be critical when the model is applied to ostensibly altruistic behavior.

Either such behavior is instrumental—that it represents reciprocal altruism—and is not truly moral, or the risk associated with the action taken is considered worth the effort in attempting to achieve a desired outcome. Principles of the model employed here are applicable to all classes of behavior, including behavioral *incidents*, behavior *sequences*, and behavior *patterns*.[31] The model applies to each behavioral type as follows:

- **Intrinsic behavioral incident**
  OA desire/need strength (netwant)
  OC cost elements (netcost)
  (+) area: The domain of netwant/netcost factors where the dominant emotion results in behavior. Either positive or negative emotions may lead to a behavior as one anticipates continuing or eliminating a behavior, situation, or event.
  (-) area: The domain of the netwant/netcost relationship where the dominant emotion results in alternative action. It is identified here as nonbehavior in order to indicate that some considered behavior is not carried out.
- **Instrumental behavioral incident**
  OA instrumental value; need strength of total behavior sequence
  OC instrumental action which is added to all other costs, pertinent beliefs, and evaluations.
  (+) area. The domain of behavior; the dominant emotion is *at least in part* one of satisfaction.
  (-) area. The domain in which willingness is insufficient to cause a particular (instrumental) behavior.
- **Total behavior sequence**
  OA netwant
  OC netcost
  (+) area: The domain of all action, including instrumental behaviors.
  (-) area: The domain of activity directed differently from the total behavior sequence under consideration.

- **Behavior pattern**
  OA netwant
  OC netcost
  (+) area: The domain of all action, including total behavior sequences.
  (-) area: The domain of activity directed differently from the behavior pattern under consideration.

Willing

The notion of *willing* as it identifies a motivating element refers to the acceptance of costs in the interest of gratifying some desire.[32] When individuals carry out actions that they are only "willing" to perform, the proper interpretation is that the behavior under consideration is, in fact, instrumental. It is a *cost* as related to some more compelling urge. Failure to appreciate the special status of this affective experience has consistently resulted in a misinterpretation of behavior. The girl who studies, although she would prefer to play and the young man who works at a distasteful task each provide evidence of the existence of an affective state which appears to violate the principle that one must desire that which they pursue. If the function of emotion is to recommend behavior, it may appear that such ostensibly negative activities must be desirable; must be enjoyed for themselves.

The confusion is caused by the assumption that a particular activity always represents *intrinsic* behavior. If a behavior which would not otherwise be indulged is believed to lead toward a desired state, or if the culmination of the act does in fact lead in such a direction, the anticipatory emotion (i.e., *willing*) is quite distinct—the opposite of that of *desiring*. Willing specifically refers to behavior that one does not desire to undertake for its own sake; to instrumental behavior. Dieting, in order to lose weight is usually considered to be an uncomfortable experience. However, if individuals avoid eating for that purpose, their behavior must be interpreted in terms of the longer range goal. Any positive feeling results from their having arrived at a decision to behave in a specific manner. They are *willing* to accept the discomfort or suffering involved.

Such behaviors represent costs to the total sequence. When altruistic behavior is under consideration, the sociobiological interpretation is that individuals are willing to sacrifice now for a later reward (reciprocal altruism), or because of an urge to conserve their lineage (kin selection or inclusive fitness). But much self-sacrificial behavior is based on the self-transcendent desire which is essential to the enhancement of a gene pool rather than to the welfare of the individual.

Critical Behavior Characteristics

Several aspects of behavior which will be employed to provide evidence of the existence of genuine altruism can be identified. Each of these concepts must be taken into account when a behavioral episode is being analyzed.

- *All behavior is outcome oriented.* In many instances, it is appropriate to identify a behavior as intrinsic, in that it is performed solely because the individual wishes to enjoy the experience. However, the reason that such enjoyment has evolved is because of its potential for resulting in an individual—and thus gene pool enhancing—outcome. This factor is somehow overlooked by those who make social judgments. It has been argued, for example, that sexual behavior is enjoyable because one of its purposes is to provide enjoyment. It may also be assumed that the experience of observing a sunset or a painting has no purpose beyond pure pleasure. A more accurate interpretation would be that such experiences, originally functioning to encourage action, have evolved to the point that they are indulged for the stimulation they provide—much as many foods without nourishment value are eaten solely for their taste.

It is not difficult to appreciate that in the case of self-assertive and self-protective needs, outcomes are critical. Though food tastes good, the development of a healthy individual is the basis for that taste. If altruism is to be appreciated as being an essential characteristic of human behavior it must be recognized that the consequence rather than one's motive is the defining feature in that case as well. Philosophers have, however, in many instances, taken the position that moral evaluation is properly assigned to the *reasons* for which one behaves. It is what one chooses to do, they contend, rather than what eventuates that determines the moral status of an action.[33] Altruistic behavior would be absurd unless there was potential profit for someone involved. "Good intentions" have, in fact, been criticized when desired results have not been achieved. In the case of morally relevant situations, those results are related to survival/growth or fitness factors, both for the behaving individual and for the human gene pool.

Outcomes are considered critical in such philosophies as *act* utilitarianism, which is based on the contention that actions are morally correct to the extent that they promote happiness, (i.e., a happy state of affairs—a *consequence* of morally positive activity). In point of fact, act utilitarianists have been criticized for just that; for analyzing morality "entirely through actions and their consequences, never through motives or intentions."[34] Hull complained that "according to the biological use of these terms [altruism and selfishness] intent is irrelevant; all that matters is the effect."[35] His interpretation is correct. Behaviors may be considered to have a moral component to the extent that they facilitate particular results. The moral sense provides incentives that encourage altruistic behavior; that create a *willingness* on the part of individuals to accept the cost associated with generous or self-sacrificial action.

An argument for rewarding positively motivated beneficent behavior (and for punishing that which is based on selfish motives) can be made on the basis of the inefficacy of a genetic plan that was dependent on the accident of the occurrence of positive outcomes and on the advantage of rewarding actions that are *most likely* to lead to desired consequence. It would be extremely inefficient to provide rewards only at the point of consequences. Furthermore, a program of rewards that causes individuals to perform any deliberate action is based on rewarding what one *does*, rather than on what eventuates. Thus a child is encouraged to avoid danger, etc., though the outcome—to be safe—is the goal. The motive for such behavior resides in the genetic mechanism, where the avoidance of danger is experienced as a positive signal (i. e., a need).

Sociobiological interpretations of behavior are attacked by those who appeal to intentions and the decision making process. Solomon, on reviewing the results of a seminar on morality and sociobiology, said "our first criticism of the sociobiological concept of altruism [is] that it is defined in terms of consequences."[36] Kowalski pointed out that since moral behavior presupposes a possibility of alternative actions, "How can [moral] states be explained by the sociobiological approach?"[37]

His argument was that since "morality is subject to a tremendous cultural elaboration of prohibitions and rules," outcomes or consequences can be only of secondary consideration.[38] For each of these moral philosophers, individuals are assumed to be subject to moral sanction on the basis of whether they are motivated to act altruistically. The behavior, not the outcome, is believed to be critical. But in this instance, the sociobiologists are correct. Outcomes are the goal of *all* behavior.

- *Costs represent the sacrifice of alternative needs or desires.* As any behavior (A) is contemplated, other behaviors as well as other needs or desires represent counterbalances. When a competing behavior (B) comes into focus, (A) contributes to the decision as a negative, or cost, element. If a putative cost element has no application to any other possible behavior, it is not, in fact, a cost. This is true of any deliberation, including, for example, the consideration of a palpably negative act. If one is considering whether to rob a bank, the behavior must provide some positive anticipatory emotion. In considering factors associated with refraining from such a robbery the potential pecuniary gain is experienced as a negative factor. The strength of the valence is determined by attitude, opportunity, and other factors about which judgments must be made.

- *All behavior is fully determined by the netwant/netcost ratio.* Given the belief that one has the requisite capacity and opportunity, and when it is believed that the potential reward associated with an anticipated behavior is greater than the cost involved, the individual *must* behave. Under those circumstances in which it is believed that the cost of such behavior is greater than the projected gain, *no behavior is possible.*[39] The reason for this is that all costs represent competing wants. If one has a stronger desire to perform an alternative behavior, and costs—including the sacrifice of the behavior under consideration—are acceptable, they must choose the alternative.

The determinate aspect of the process does not absolve a behaver of responsibility since the scale on which wants and costs are weighed is governed by the interests of the individual. Want/cost relationships determine what the individual *does*, but they are a consequence of what the individual *decides*, and decisions are based on the relative value of alternative behavioral options.

- *All behavior is based on belief.* It was pointed out earlier that mental knowledge is experienced as a matter of belief, which falls inescapably short of certainty. Thus, when an action is contemplated cost/gain factors are never more than "best estimates" in spite of what often seem to be incontrovertible facts. If a person attempts some feat, e.g., to jump over a hurdle, some belief must be involved. The belief is not, however, necessarily that the attempt will meet with success. It may be, for example, that the individual believes someone will be impressed with the effort, or that some negative sanction can be avoided. The critical factor is that when a behavior is considered beliefs must always be involved. It is the interaction of such beliefs rather than any collection of "facts" that makes the determination.

- *Wants and costs are totally independent of each other.* If one desires to eat a particular food, the cost may influence the decision to purchase that meal, (*a behavior*), but it will not increase or decrease one's desire for it. Similarly, alterations in the strength of a desire have no influence on beliefs regarding the cost involved. Although a smoker's desire for a cigarette probably increases as a function of time the risk (a cost) involved in smoking in a prohibited area remains the same.[40]

Reason, Morality, and Altruistic Behavior

Many philosophers have appreciated the fact that a distinction must be drawn between the roles of morality and reason as behaviors are considered. Hume, for example, argued that "the sole purpose [of reason] is to compare ideas; it is powerless to distinguish between the good and evil in action."[41] Furthermore, he distrusted reason as an evaluative mechanism. Among other things he said that "[reason] is slow in its operations; appears not, in any degree, during the first years of infancy, and at best, is in every age and period of human life, extremely liable to error and mistake."[42]

First, of course, reason is *not* necessarily slow. It is sometimes incredibly rapid, and it *does* appear in early infancy. More important, however, is his claim that reason is "liable to error and mistake." This comment reveals a misunderstanding of evaluative processes that destroys his thesis. To appreciate the conundrum, consider Hume's assertion that moral (ought) statements cannot be based on reason.

> In every system of morality which I have hitherto met with...the author proceeds for some time in the ordinary way of reasoning...when... instead of the usual copulation of propositions is and is not, I meet with no proposition that is not connected with an *ought* or an *ought not*.[43]

Hume argued that such authors are obligated to explain the basis for such statements, saying that if they would do so it would become obvious that "the distinction of vice and virtue is not founded merely on the relation of objects, nor is perceived by reason."[44] He claimed that moral connections are made on the basis of a moral sense or a "moral sentiment" which, though this was an accurate analysis, and is consistent with the contention of this text, unfortunately did not solve the problem. What it did was to introduce one more evaluative element which was hardly what Hume was looking for (and which he did not, in fact, ultimately support).

In point of fact, Hume's views on the concepts of social justice and "public utility" do not bear out his claim that reason and morality are separate concepts. He asked *"what must become of the world if ...[unjust] practices prevail. How could society subsist under such disorders?"*[45] His question calls for a response based on reason. Since it would be unreasonable, he implied, to expect an unjust society to subsist, individuals should be just. And here is where Hume, and all philosophers who take a similar stance, founder.

The obvious response to Hume's question is that it *would* be reasonable to assume that a just society could be expected to meet most people's needs. One problem with such an argument is, however, that it represents an amoral—perhaps an *immoral*—proposal. To do that which meets everyone's needs is probably as ultimately selfish as any other form of behavior. To do what is right because the individual (along with others) will profit is hardly altruistic. There is, however, a far more serious error.

To apply reason to a problem in the abstract is an entirely different matter than applying it to the deliberative process. It is obviously reasonable to follow rules of moderation if one wishes to enjoy good health. However, when one is considering whether to eat a second dessert the taste value is weighed against costs in health, appearance, social acceptance, and other factors. While it may seem reasonable to consider a society better in which citizens act altruistically it is not reasonable—it is, in fact, *unreasonable*—for individuals to sacrifice for others, unless they can expect to gain more than is given, or unless there is a willingness to perform such behavior as a beneficent act, as will be demonstrated.[46]

The confusion caused by this misinterpretation of the roles of reason and morality has led to many strange philosophic claims. It is said, for example, by existential or phenomenological psychologists, that people really "want to be good," but are prohibited by the influence of social forces. At the other extreme are claims that some individuals (e.g., serial killers), desire to be evil. However, all people act in order to meet whatever desire is involved, with some behavior being subject to evaluation on the moral scale. The same can be said for the assumption that people desire to be reasonable. So long as such interpretations of behavior are employed, the true motives for action will not be discerned.

Hume's contention that reason is subject to "error and mistake" is not supportable because no affect can ever be shown to be right or wrong. Such claims as that a particular dish of ice cream is "delicious" or that an attacking tiger is "dangerous" are not subject to a truth test. Hume recognized the limitation of reason in his analysis of the relationship between cause and effect with his conclusion being that belief about such a necessary connection is "the foundation of our inference from one to the other

[and] is the transition arising from the accustomed union.⁴⁷ However, he did not seem to appreciate that his description of the reasoning process differed not at all from that of the moral process. In each instance, one *believes*, but can never have precise knowledge.

How do questions such as: "Should I invest in the stock market today?" or, "Should I marry that person?" or even, "Should I cross the street now?" differ in principle from "Should I assist the disabled?" or, "Should I risk my life for my family?" In each instance, no reasonable—correct— answer can be provided. Why, then, is a special case made of the moral aspect of the "ought" issue when in point of fact the "should" question is applicable equally often to situations with no moral application at all? Consider once again Hume's famous dictum, this time substituting the concept of reason for that of morality:

> In every system of *reason* which I have hitherto met with...the author proceeds for some time in the ordinary way of thinking...when...instead of the usual copulation of propositions *is* and *is not*, I meet with no proposition that is not connected with some notion of *ought* or an *ought not* where ought refers to the success probability of some behavior.

The principle involved is that moral evaluation, like that of reasoning, performs an essential function in the decision making process. How a behavior is evaluated on the moral scale is, of course, subject to the influence of learning. In its pristine state it is experienced as an affective response to certain occurrences, (e.g., wanton slaughter) just as is the loathing of certain experiences such as the touch of slime, or the sight of a snake or spider. But individuals can—and do—learn to alter reactions to various stimuli. The same is true of one's moral viewpoint regarding specific ethically charged behaviors. What is considered ethical in one situation may not be so viewed in another.

It was pointed out above that the increased capacity for rational behavior is assumed by some philosophers to provide for the emergence of what may represent genuine altruism. As individuals—or groups—have become more intelligent, they are believed to have developed the capacity for acting more compassionately; to empathize with their fellow creatures, and at the human level, to respect all forms of life. However, although a moral sense has characterized the evolution of the human species, the capacity to employ reason has been the most significant cause of the movement from mutuality to the "look-out-for-number-one" philosophy that has been preached by a variety of philosophers and social activists, and practiced by those in political power, from the earliest stages of civilization.

Although putatively altruistic behavior may be evaluated on the scales of both morality and reason, the scales measure totally different factors. Reason speaks to the probability of a salubrious outcome, morality is concerned with the propriety of a behavior. "Cooperative," behavior, being considered reasonable, is practiced solely in the anticipation of reciprocity. Altruistic behavior, by contrast, may be characterized by both; by morality in terms of the goal of helping others, and by reason as to the relationship between gain and cost in pursuing beneficent behavior.

Feeling states are inevitably related to percepts in some way that potentially lead to profitable activity, and ultimately to optimal gene pool situations. Morality and reason must, therefore, meet that requirement. The principle that is of concern in this text is that *each evaluative scale serves a unique function*. Altruistic behavior is based on the influence of one of them. Sociobiologists have been unable to explain this "bizarre" behavior. It appears to arise spontaneously—on the basis of a (moral) sense that prescribes "shoulds" and "oughts," that most philosophers and biologists insist could not be generated by the activity of selfish genes.

Summary

In this chapter the elements of a motivational sequence were described. The purpose was to demonstrate that all behavior (i.e., deliberate action) is dependent on the animation of some desire. The activation of desires is dependent on the status, or operational capacity of many mental elements, including that identified as *self-transcendent*, because it represents a desire to see the welfare of others enhanced or safeguarded.

Critical to the evolution of deliberate action was the emergence of *mind* which has resulted in a number of changes in response patterns that are relevant to the problem of altruism. Most significantly, it has become possible for the individual to delay responses to internal (drive) and external (environmental) stimuli. Such response delays are characteristics only of a state of consciousness. The rigid demand for a response that characterizes mindless creatures, on the other hand, prohibits any choice between potential activities. Belief, with its constraints on sure knowledge comes into play.

Intelligence, which may be defined as "*the potential for change that is expressible through learning,*"[48] is a characteristic of all entities. An emergent form of intelligence is related to thought processes. Such a skill provides opportunities for actions that preferentially reward those most intellectually advantaged. However, it carries with it the potential for excessive egoism. What is needed is some technique for monitoring such behavior which, left untrammeled, is responsible for potentially grave damage. The moral sense provides such a mechanism.

Desires represent a form of *instinctive* knowledge. Although they are positive aspects of the mental process, they are experienced as negative feelings. Thus, both selfish and altruistic motives may be considered signals to take action that may be ameliorative. The distinction to be drawn is between behaviors that are ultimately related to serving the immediate or inclusive self, and those that are directed toward the more general welfare. In the latter instance—where the individual must endure some actual or potential sacrifice—an affect defined as "willing," is involved. Altruistic behavior may, thus, be defined as activity which is performed willingly in response to the self-transcendent urge and is evaluated positively by the moral sense.

The rational and moral senses provide rudders or "device[s] for governing, directing, or guiding a course." [49] The course involved is the twists and turns of the motivational process. The function of such stabilizers is to influence behavior by providing scales against which to evaluate an action. The rational scale provides data that bear on the success probability of a contemplated behavior. The moral scale evaluates behavior as to the life enhancing aspect of its consequences. Genuinely altruistic behavior cannot meet the test of rationality if the welfare of a behaving individual, or that person's kin are the primary concern. Only if some community to which the behaver feels an allegiance is positively affected can altruistic behavior be considered reasonable. The moral sense, thus, provides a bridge between individuals and their gene pools.

Emotions, which are peculiar to living creatures—and among the living, perhaps only to those that have attained a high level of neurological sophistication—perform the function of providing a forum for the consideration of alternative behaviors. In the most highly developed forms of life, the ability to recall and to imagine make possible emotional reactions to recollections and hypothetical situations. Every emotion, from joy to distress, from happiness to anger, from pride to guilt, is designed to provide an impetus to action—though in many instances (e.g., grief) no behavior is possible.

Both the rational and moral senses are subject to distortion when emotional levels are high. The frustration of efforts to obtain food may cause an individual to believe that an objectively (?) irrational action may solve the problem. Similarly, the impact of an unprovoked assault by a stranger may cause one to attempt to justify an extreme response. This ostensible alteration of the sequence from reason and morality to emotion merely points up the fluctuating impact of each element in the motivational chain. Since emotions are functional, the issue that must be dealt with is: Why are positive emotions so often experienced when individuals behave in ways that do not serve their own interests?

Behaviors may be classified as intrinsic—those in which the experiences are desired in themselves—or instrumental—those that represent cost factors as they are related to goals for which the individual is willing to pay a price. All behaviors are fixed by the netwant/netcost ratio, and by beliefs about potential outcomes. Costs represent limits on the opportunity to express other needs or desires. In order to understand altruistic behavior the self-sacrifice involved provides evidence of the fact that some greater goal is involved. The principle of reciprocal altruism (that certainly describes some behavior) is incompetent to explain all other-directed action. The moral sense can be shown to represent a vehicle for the positive evaluation of genuine altruism—behavior based on the transcendent desire. Unfortunately, a significant number of ethologists and others of the biological class have come to deny that any such sense exists, and that, thus, genuinely altruistic behavior could not occur.

Altruism and the Motivational Process II

Chapter 4 Notes

1. There are obviously a number of characteristics of individuals, such as the existence of two kidneys that seem difficult to explain. However, the general principle cannot be evaded on the basis of a few, as yet unexplained exceptions.
2. Functional characteristics, of course, did not *emerge* because they were advantageous, but did *evolve* for that reason. The distinction is critical to an appreciation of the holarchic model.
3. Although some zoologists, on observing the hydra occasionally contracting into a ball shape, assume that it is "sampling" its environment, it should be clear that the process is as automatic as any other.
4. This type of evolutionarily stable strategy, said Dawkins, "enables us...to see clearly how a collection of independent selfish entities can come to resemble a single organized whole," (1976, p. 90). He contended that what looks to be beneficent in this type of behavior is, in fact, self-serving. In this instance "selfish" behavior is rewarded, and explanations that refer to "profit to the gene pool, or the species," can be successfully screened off. (An opposite interpretation is, however, equally feasible.) Although symbionts assist each other, unlike parasites, which usually injure their hosts, neither altruism nor selfishness is involved. In spite of the fact that an individual may alter its course of action, or modify its shape, such options are extremely limited. No affect can be discerned.
5. Wonderly (1991), p. 153
6. Carty demonstrated that certain caterpillars (*Lymantria monacha*) will approach a series of black stripes painted on a wall. He stated that "if four stripes are equally spaced around the wall, caterpillars...will go in approximately equal numbers to all of the stripes.... [However] when the stripes are of different widths more caterpillars go to the widest," (1971, p. 128). Snails, he said, "usually will travel in straight lines...[And] carapid beetles...usually behave phototactically," (*Ibid.* p. 131). In these examples such terms as "usually," "most," "more," etc. indicate that a variety of responses occur, depending on environmental circumstances. The information that the organism receives is interrupted in some way.
7. Wilson reported that "some arachnids [a type of spider]...carry their newly hatched young around in brood pouches on the abdomen," (1975, p. 336). He contended that the behavior of the parent wasp is evidence "for regarding the complexity of adult-offspring relations as a true measure of social evolution," (*Ibid.* p. 345). Such behavior inevitably puts individuals that protect others at some cost or risk, even to their own lives.

8. Wonderly (1991), p. 522. While many processes occur at a *nonconscious* level, it absurd to consider them "mental."
9. This is exemplified in the mental processing, and the contemplated behavior of a hungry person. How hungry am I? Are the costs associated with eating worth the gain? And, at least for humans, what are the moral factors associated with a particular behavior?
10. The notion that cognition represents a class of action is not in conflict with the neo-behaviorist contention that thinking is, in fact, implicit or unobservable behavior. However, the behavior involved may or may not be the subliminal physiological response that is claimed. To think, to consider, or to measure, all represent processes that describe actions rather than actors, and their operation as neural firings or molecular reorganizations represent only their mode of operation. Cognition, thus, describes only the *action* involved in mental processing. It refers to what one *does* by way of manipulating data and deriving relevant concepts.
11. Stroud (1977), p. 74
12. Those who attempt to distinguish belief from knowledge contend that out of the affective experience, true knowledge may develop. Russell said that "every case of knowledge is a case of true belief, but not vice versa," (1948, p. 154). And further: "What an asserted sentence expresses is a *belief;* what makes it true or false is a *fact* which is in general distinct from a belief. (The sun is a fact; Caesar crossing the Rubicon was a fact,)" (*Ibid*. p. 143).

   Ayer insisted that to know something is to *believe* it to be true, that it be true, and that there are good reasons to believe it. He distinguished belief from knowledge by saying that one may be completely sure of a belief that does not meet the condition of being knowledge (being in fact false, for example). "But whereas it is possible to believe what one is not completely sure of...this does not apply to knowledge," (1956, p. 16). Ryle contended that although both terms refer to dispositions, "knowing refers to a capacity and believing to a tendency" (1949, p. 134), adding that knowing is "to be equipped to get something right," (*Ibid*.). This suggests a correspondence between the knower and the "truth out there." The absurdity of these and similar claims is discussed in detail in the text, *Motivation, Behavior, and Emotional Health,* (Wonderly, 1991).
13. The term "drive" has, in fact, become anathematic to philosophers, many of whom consider it no more than "the reification of a motive," (McShea, 1990, p. 85). Biologists become quite exercised about its use, proposing that since drives "have been driven from the study of comparative animal behavior...it is time we evicted them from human psychology," (Ibid.).

McShea complains that "a few recent writers...offer shopping lists of human needs and wants beginning with food and water.... (*Ibid.* p. 87) This distresses him, he says, because in truth "it is the particular feelings that urge [people] on, not a hypothetical drive," (*Ibid.* p. 85). One wonders what he considers to be the source of such "feelings."

14. White (1959), p. 300
15. The response to the proposition that the anticipation of a fine meal or an exciting sexual experience represent positive desires is that the experience in each case actually represents the anticipation of attaining a need. The test would be to eliminate the possibility that either need could be obtained. The negative nature of the desire should be quickly appreciated.
16. Wonderly (1991), p. 234
17. Hume quoted in Aiken (1970), p. 33
18. Criticizing the tendency of men to attach themselves to both worthy and unworthy causes, Koestler said, "*the egotism of the group feeds on the altruism of its members*," (1978, p. 83). Smythies spoke of the "divorce between the two parts of the brain, the rational and emotional," (1965, p. 276). And Price argued that moral ideas are no more than matters of "understanding" which he contended are based on reason. Clarke added that "the absolute and immutable character of moral distinctions is such that they can be known only by reason. Therefore the moral faculty could not possibly be a sense," (Clarke, referenced in Sprague, 1967, p. 387).
19. Wonderly (1991), p. 238. The latter scale has been defined as the moral sense; the mechanism which "binds [each] individual...to other individuals, and to life in general," (*Ibid.*).
20. Spinoza contended that "men who are governed by reason...desire nothing for themselves which they do not desire for other men, and that, therefore, they are just, faithful, and honorable," (1963, p. 172). But either such individuals are foolish, (since they risk the loss of personal profit) or assume that such behavior will ultimately result in advantage to themselves. They could not be guided by a rational sense that spontaneously put the welfare of others above their own.

    Wilson argued that as life becomes more sophisticated altruism declines. He suggested that there is a negative relationship between intelligence and altruism, which contradicts the view of those like Spinoza who contend that morality and reason develop in tandem; that increased intelligence will inevitably lead to an increase in virtuous behavior.
21. Albert & Denise (1988), p. 3. In view of the fact that reason, as it relates to behavior represents a capacity to discern the probability that an action will result in a particular outcome, there are many situations in which it

may *appear* that reason runs parallel to morality. One of these would be in the consideration of whether an action will result in a morally positive situation. What might one person *do* to assist another? Which of several possible courses of action is most likely to maximize the happiness of another? Here the cognitive process is employed in the making of a determination. It does not legitimize the notion that it is reasonable to be moral–*–which it is not*. What it does indicate is that if one wishes to act in a certain way on the basis of some desire, a particular procedure may be considered as to its proficiency in meeting that condition.

22. "The role of both the moral and rational systems is judgmental.... They are involved as estimators of the quality of potential [or actual] activity.... Neither is a form of desire. They follow and judge potential or actual situations and events which are believed to have moral and/or rational components." (Wonderly, 1991, p. 239). Warnock agreed with the position taken here, that moral and rational judgments "are all instances of *appraisal*, or *evaluation*, and indeed of *practical* appraisal, since we are dealing essentially with judgment of human conduct," (1971, p. 9).

23. Williams correctly pointed out that "no biologist believes that there is a gene locus with some ordained role in adjusting altruism," (1989, p. 188), to which Stebbins added, "such phrases as 'genes for altruism' and 'genes for spite' give a misleading impression to readers who are not fully acquainted with modern biology," (1982, p. 382). However, altruism does, in fact, have a genetic *base*.

24. McShea, for example, said that "Every slightest voluntary motion, every desire or aversion...is the expression of one or more emotions" (1990, p. 15). Most authorities agree.

25. Consider the reaction of an individual who has just eaten all that is wanted, if additional food is proffered.

26. Portman (1949), p. 16

27. Guisinger & Blatt (1994), p. 104

28. *The Random House Dictionary of the English Language* (1987)

29. Franklin (1758)

30. Wonderly (1991), p. 449

31. An *incident* is a single behavioral event (e.g., kicking a ball), a *sequence* includes all elements of a series of incidents (e.g., playing a game), and a *pattern* refers to a "typical but not continuous action performed to meet various desires," (Wonderly, 1991, p. 383). Thus, contributing to a charity on a regular basis may be said to be a behavior pattern.

32. Harriman, for example, proposed that "will" is "a broad concept... designating persistence in voluntary choice or a direction of behavior toward

remote, highly valued goals," (1974, p. 219). The term has been defined, in many ways. First as "the faculty of conscious and especially deliberate action," (*The Random House Dictionary of the English Language*, 1987, p. 2174). As such, that definition is offensive only to those philosophers that deny the dualism it suggests. However, that is followed by the "power of choosing one's actions: *to have a strong or a weak will," (Ibid.*). As it stands, this seems to describe a motivational element—a capacity to act.

In each case, however, the critical aspect is ignored. A clear example is found in Flew's *Dictionary of Philosophy*, where the "weakness of the will" is described. Flew contends that the term refers to situations in which an individual goes against his better judgment. "It is usually assumed that an agent intentionally performs some action *x* rather than action *y* only if he wants to do *x* more than he wants to do *y* ; and furthermore, that if he thinks, all things considered, that it would be better to do *x* than *y*, then he 'wants' to do *x* more than he wants to do *y*," (1979, p. 372). He fails to recognize the appropriate interpretation of the term as it refers to motive. Willing behavior is always instrumental. If one does *x* it may represent a price paid for attaining some other goal.

33. Wolff suggested that the moral evaluation of a behavior requires that "intention, deliberate choice among equally determined actions, and awareness of the social consequences of alternative actions can be assumed," (1980, p. 84). Kowalski argued that "morality is characterized by intentional states," (1980, p. 233). And Stent held that "whatever else morals may be...the concept of morality pertains to an *intentional state* of an agent," (1980, p. 17). He pointed out that attempts to consider morality as related to the consequences of an act, rather than to its motive, separate the concept from the "morals of ordinary discourse," (*Ibid.* p. 24).

34. Flew (1979), p. 360
35. Hull (1989), p. 247
36. Solomon (1980), p. 265
37. Kowalski (1980), p. 232
38. *Ibid.*
39. Consider a situation that suggests that one is free to take action in a given situation in spite of the fact that it is believed that costs are higher than potential gain. In providing a reason for violating the principle, one is reduced to acknowledging previously unconsidered positive factors, to having overestimated costs, or to invoking some *causa sui*.
40. There is a strong appeal to the idea that what is unavailable because of the cost involved, becomes less desired. The "sour grapes" principle developed by Festinger in his cognitive dissonance theory suggests that costs

do alter need values. What happens, however, is that the individual either suppresses the desire for a particular item or experience, or replaces it with another. In the latter instance, what happens is that one behavior (or need) replaces another. Desires, *per se*, are independent of costs—which should be clear when it is understood that costs are, in fact, competing needs or desires.

41. Hume (1956), p. 383
42. Hume quoted in Aiken (1970), p. 43
43. *Ibid.*
44. Aiken, in a vain rescue attempt, made the claim that for Hume, "it is only when a character or act is considered in relation to a certain kind of *feeling* or *sentiment* that we properly 'denominate' it as morally good or evil," (1970, p. 43). And that is the crux of the matter.
45. *Ibid.* p. 201
46. One of the possible causes of confusion regarding the relationship between reason and morality may stem from the erroneous notion that reason is not a motivator while morality is. "Reason," said Hume, "is no motive to action, and directs only the impulse received from appetite or inclination," Hume quoted in (Aiken, 1970, p. 269), and "when it excites a passion by informing us of the existence of something which is a proper object of it," (*Ibid.* p. 34). However, he concluded, "it is not pretended that a judgment of this kind, either in its truth or falsehood, is attended with virtue or vice," (*Ibid.* p. 37).

    Hume contended that "to have a sense of virtue is nothing but to *feel* a satisfaction of a particular kind," Hume quoted in Aiken, 1970, p. 44). Morality, he said, " is the object of feeling, not of reason," (*Ibid.* p. 42). While his reference to feeling is accurate as regards morality, it is equally valid as it refers to reason, and to aesthetics as well. All such scales function to improve the ability of the individual to respond effectively to internal as well as external stimuli. In no case can an evaluative mechanism provide a precise, or "correct" answer to a behavioral determination.
47. Hume quoted in Edwards (1967), p. 80
48. Wonderly (1991), p. 148
49. *The Random House Dictionary of the English Language* (1987), p. 1679.

Chapter Five

The Sociobiological Thesis

> A science cannot be played with. If an hypothesis is advanced that obviously brings into a direct sequence of cause and effect all the phenomena of human history, we must accept it, and if we accept, must teach it.
>
> *Henry Adams*

In spite of the analysis of behavior proposed in Chapter 4, sociobiologists have proposed an explanation for the problem of putatively altruistic behavior by reducing all behavior to actions that serve the behaving individual or that person's kin, or is performed in the anticipation of receiving some future recompense. The basis of their conviction is simple and straightforward. Mindless genes are most likely to be represented in ensuing generations if they survive and generate offspring. Not because of any purpose, goal, intention, desire or other motive that appears to characterize more complex forms of life (e.g., humans), but simply because one condition (reproductive fitness) is essential to the occurrence of another (representation in the following generation). Geneticists and molecular biologists have provided powerful evidence to support that view.

Many scientists have, however, questioned their conclusions. Stebbins, for example, pointed out that "one of the biggest gaps in modern biological knowledge is our almost total ignorance of the mechanisms by which disparate programs of differential gene action are brought about."[1] Other problems come from those who ask: Why is it appropriate to accept the principle that cells function cooperatively, that individual genes work in concert, and, thus, in the interest of more complex existents, while denying that the infrastructure of existents such as humans may include instructions that recommend obligation to individuals and organizations that are external to the physical body of the individual involved?

Sociobiologists are not deterred. They insist that what appear to be generous or self-sacrificial behaviors are, in fact, disguised self-enhancing actions, although no intention to deceive need be involved. Not only is the whole process simplified, but any ostensible exception can be explained by appealing to principles that biologists insist must be operating.[2] The conclusion that the subsistence of each individual's lineage is involved in the genetic process shall not be questioned here. However the refusal to accept the possibility that true altruistic behavior occurs shall be challenged in spite of the fact that sociobiological tenets have, in general, much to recommend them. To understand the reason for this deference, the lynch pins that hold the complicated edifice that is sociobiology together must be examined.

From the time that Dawkins' 1976 book, *The Selfish Gene*, was published, there has been a steady increase in the acceptance of the thesis that there is a genetic basis for moral sentiments. However, sociobiologists want more. MacDonald reflected the conviction of many biologists that "the self-interested side of the implications of evolutionary theory...has not received the attention it deserves within developmental society."[3] If genes control behavior, say sociobiologists, such action must be geared to the survival potential of their progeny. Wilson argued that when a bird gives a warning call that ostensibly aids others, its behavior can be just as appropriately interpreted as being based on selfish concerns. The emotion involved is based on the desire of the bird to save itself, an awareness of potential danger, and a technique for escaping from harm. He quoted Maynard Smith as having said:

> Warning calls are fixed by kin selection and sustained outside the breeding season in evolutionary time owing to the probability that close kin are near enough to be helped.[4]

Dawkins offered a similar explanation. Discussing the behavior of flocks of birds when danger is immanent, it is presumed by the laity that the alarm sounded by one of them is given in a effort to protect its fellows. However, Dawkins warned, such behavior may be equally well explained as directed toward the safety of the individual making the call. If a bird who sees a hawk stays where it is, even remaining quiet, it is more at risk than if it flies into a nearby tree. But if it were to fly away alone it would be more likely to be attacked since predators prefer to attack animals that are separated from their group. What might the imperiled bird do? "The best policy is to fly up into a tree, *but to make sure everybody else does too.*"[5] Such an explanation, which reveals the innate selfishness of the bird, solves the problem to Dawkins' satisfaction. Other sociobiologists have provided examples of the same type. Williams contended that "The gene pool of a population is a record of reproductive success and failure in that population, and at conception an organism gets a *sample* (italics added) of that record."[6]

The most difficult problem in dealing with the contentions of sociobiologists is to identify precisely what their position on altruism is. In their various writings, they have sometimes denied the possibility of truly altruistic behavior, sometimes "allowed" that it happens in limited circumstance, and at other times have avoided the issue as if in so doing they may not have to deal with the controversy. Dawkins, for example, regarding the issue of genuine, disinterested, true altruism said at one point, "I am not going to argue the case one way or the other."[7]

The fact is that if sociobiologists allow that *any* truly altruistic behavior occurs they have lost their argument. They have not accounted for a behavioral influence—an evaluative mechanism—which under some circumstances calls for individuals to sacrifice themselves without hope of indemnification. There is, of course, considerable allusion to a moral sense in professional writings but such references are not well publicized. Kagan for example, said that without a capacity that "nineteenth century observers called a *moral sense*"[8] children could not be socialized. What label would be applied by twentieth century biologists to such a motivational factor ?

Wade stated the currently accepted view that "any group that contains even a single non-altruist will be converted by individual selection within the group to a...non-altruistic group. Any migration between altruistic and non-altruistic groups...has disastrous consequences for the evolution of altruism."[9] Ridley made the same case. "If a group of altruists ever did come into existence organismal selection would rapidly convert the group to selfishness."[10] But would it? Would the appearance of an altruist represent a harmful mutation? *Are people absolutely selfish? In all situations? At all times?* If the answer is yes, Dawkins and other sociobiologists can appeal to no aspect of their motivational apparatus to explain why (genuine) altruistic behavior should be encouraged. Not now. Not tomorrow. *Not ever*. The task, therefore, is first, to determine to what extent the interpretations of altruistic behavior provided by sociobiologists are convincing.[11]

The Sociobiological Argument

Kin selection or inclusive fitness

These two terms have been employed to describe aspects of the same general process. Mayr defines kin selection as "selection for the shared components of the genotype in individuals related by common descent."[12] It is, thus, a technique by which the degree of genetic relationship between a donor and a donee is taken into account as one considers performing an unselfish act. He defines inclusive fitness as "the sum of an individual's own fitness plus all of its influence on fitness in its relatives other than direct descendants."[13] For the purpose of analysis, the distinction is probably inconsequential, since both refer to a class of behavior that is influenced by a willingness to risk or sacrifice to assist others with some degree of genetic relatedness.

Nineteenth century biologists had hinted at a biological basis for social behavior long before evolutionary principles were fully developed. McShea quotes Hobbes as making a case for kin preference in saying that "the survival at which the passions are directed is always the survival of the particular passionate individual, except in the very few cases where the individuals attachment to or identification with some other persons...is so great that he or she may jeopardize his or her own interests in their behalf."[14]

Darwin, however, provided a solid scientific basis for the principle, employing it in his effort to explain the behavior of social insects, and Haldane (anticipating Hamilton's interpretation), remarked at one time that he would sacrifice his life "for three brothers or nine cousins."[15] However, Hamilton is credited with formally introducing the concept in two articles published in 1964. The factors of concern here are both the explanatory power of his hypothesis, and the extent to which the model explains how kin selection is accomplished.

In the first of his articles, Hamilton made the point that inclusive fitness "implies a limited restraint on selfish competitive behavior and the possibility of limited self-sacrifice."[16] This "limited restraint" as he defined it refers only to sharing with kin. Using the term "giving" as equivalent to "sacrificing," he said, "if a giving-trait is in question... genes which restrict giving to the nearest relative...tend to be favored."[17] He went further. "With giving traits it is more reasonable to suppose that if it is the nature of the prerequisite to be transferable, the individual can give away whatever fraction of his own property that his *instincts* (italics added) incline him to."[18]

He summarized the section by saying, "from a gene's point of view it is worthwhile to deprive a large number of distant relatives in order to extract a small reproductive advantage."[19] At another point he said: "We reject the pseudo-explanations based on the 'benefit to the species'."[20] If one rejects the species as a target, what alternatives are available? Obviously, for Hamilton, either the individual or the gene line it manifests.

Smith spoke of *absolute* fitness which, he said, refers to "the expected number of offspring contributed by (any) individuals in the next generation...[i.e.], the probability that (any) zygotes will survive to breed X expected numbers of offspring, given that [they] survive."[21] He believed, however, that the concept refers neither to genes or populations. It is rather "a property of classes of individuals and not genes."[22]

Li and Grauer propose a definition that focuses not on expected but achieved survival superiority. Fitness, they say, is "a measure of the relative survival and reproductive success of an individual or a genotype."[23] It is not what *may* occur that defines fitness, but what, *in fact*, eventuates. The difference is Sartrian. What must be attended to is not what could, but what does happen. They argue further that in view of the fact that population size is usually constrained by environmental factors it would seem more useful to think in terms of *relative* rather than *absolute* fitness in comparing competing genotypes.

Hamilton related his argument for the significance of fitness to evolutionist principles, stating that since a generalization would *seem* useful, and assuming that it *may* provide a useful summary he offered what he referred to as a "generalized unrigorous statement":

> *The social behavior of a species evolves in such a way that in each distinct behavior-evoking situation the individual will seem to value his neighbor's fitness against his own according to the coefficient of relationships appropriate to that situation.*[24]

Degler provided what he considered a *prima facie* case for the reduction of altruistic behavior to an effort to perpetuate one's gene line. "If genes play no role through natural selection, why do people care about selection at all?... Where does the idea of relation come from if not from biology?"[25] Ortner agreed, adding that, "Hamilton's hypothesis has been elaborated and utilized to formulate an integrated hypothesis relating behavior and biology for the entire animal kingdom, including man. [It] has emerged as the newly proposed discipline of sociobiology."[26] And under that thesis selfishness is the root of all behavior.

Badcock offers a typical biologist's interpretation. "If by sacrificing me, it [a gene] can secure a greater representation for itself in the future, it will have been selected for just like any other feature."[27] And Charlesworth, in describing kin selection said, it is clear that:

> animals are programmed genetically by evolution to act solely in their own interests.... Natural selection ultimately selects for one thing only; the reproductive success of genes for which organisms are nothing more than temporary packaging. Only this view of things makes sense of suicidal altruism from the point of view of natural selection without recourse to dubious concepts such as group selection.[28]

Ingold provided an interpretation of kinship theory that, in agreement with Hamilton, would allow for kin based altruism. If, by altruistic activities that are of benefit to his kin, an individual can be compensated by a gain in the fitness of close relatives, "his inclusive fitness will be enhanced, and traits conferring a tendency toward altruism will be preserved in the course of natural selection at the expense of those prescribing more egoistic behavior."[29] But how do people recognize their kin?

A variety of propositions have been put forward. Hamilton proposed that in the interest of explaining the phenomenon that individuals have discriminatory powers, perhaps "we need to postulate something like a supergene affecting (a) some perceptible feature of the organism, (b) the perception of that feature, and (c) the social response consequent upon what is perceived."[30] Holmes and Sherman proposed a similar explanation. Individuals simply match their phenotypes to those in the immediate vicinity.

> From a proximate perspective, kin recognition is usually based on spatial distribution or direct association during rearing. [There is also] an additional mechanism that could facilitate recognition between relatives who have not previously associated with each other. This mechanism, phenotype matching, is based on learning about one's own phenotype, and then comparing phenotypes of unfamiliar conspecifics to this template.[31]

Smith proposed that "one possible mechanism for kin recognition involves hypothetical recognition alleles...[which would have the capacity to] influence individuals to be altruistic toward conspecifics who manifest that trait."[32] The mechanism by which this is accomplished is presumably some genetic operation, and Segal stated that many researchers are convinced that "in humans, social bonding may be facilitated by neurological processes that underlie attraction between individuals who perceive similarities between themselves."[33]

Barash, minimizing the problem of recognition said: "It's not all that difficult. Even animals can be fairly confident that neighbors are more closely related than strangers, and genes that say 'be nice to your neighbors and less nice to strangers' are probably spread by way of kin selection."[34] Badcock introduced a Freudian element. "Both psychoanalysis and recent evolutionary biology," he said, "suggest [an] essentially *dynamic* concept...that of [an] unconscious *which is actively and involuntarily excluded from consciousness.*"[35] The assumption is that recognition may be mediated at some subterranean psychological level. Such mechanisms could, perhaps, provide the information necessary for discerning kinship levels, but they may be equally useful in relating individuals to far more general entities—e. g., demes and gene pools.

Research evidence has been provided for the purpose of validating the hypothesis that individuals are emotionally attracted to their kin and are thus apt to act preferentially toward them. Smith quotes a Porter et al. 1983 study of smell recognition in which mothers where asked to identify their own babies on the basis of the odor of shirts worn by these infants. "A significant proportion of mothers could do so, even though they had had an average of less than three hours of contact with the baby prior to testing.... Some mentioned the similarity between the baby's odor and that of other family members as an important cue."[36] In other studies, it was found that "mothers with only a few hours postnatal contact with their newborn infants were able to distinguish a photo of their child's face from photos of three unrelated infants."[37]

Twin studies offer another method for evaluating the sociobiological kin selection hypothesis. Segal says that "reunions between reared apart twins are, perhaps, human sociobiology at its finest."[38] The thrust of such studies has been that "most newly reunited MZ [monozygotic] twins, and some DZ [dizygotic] twins, establish powerful bonds *despite absence of previous familiarity.*"[39] Such pairs seem to feel that they are part of a more comprehensive existence; that they are not whole when apart; that they need their twin partner if they are to represent a total person. This provides support for what has been defined as the GST (genetic similarity theory).

In the case of the death of a twin, "given the evidence to support the bond between genetically identical twins as the strongest of human ties, sociobiological theory would predict that surviving members of MZ twinships should wail the loudest for their lost genes, louder than any other biological relative."[40] As an example, Segal quotes a surviving twin as saying "watching my twin die was the equivalent of watching myself die."[41] The data on twin pairs is extensive. The University of Chicago studied 105 pairs of twins in the 1980's. The Minnesota Center for Twin and Adoption Research is currently studying emotional results of twin loss. The consensus among researchers is that such studies provide solid support for sociobiological hypotheses.

Williams provided a defense of the position taken by Neo-Darwinist biologists (who comprise perhaps 99% of the total) regarding the basis for ethical behavior, and the implication that siblings would be valued in some hierarchical order. In making the point that mothers seem to love their older children more than their younger sibs he said: "Anyone who feels that this sort of quantification destroys the beauty of family life is a modern equivalent of those who felt that Newtonian optics destroyed the beauty of the rainbow."[42] Can that be true? Newton discovered that which *produces* the rainbow. It has been learned more recently that smog produces beautiful sunsets, cancer cells are sometimes described as "beautiful," and it is well known that the most successful prostitutes are those with the most natural and/or artificial beauty. (A surprising number of people would recommend the elimination of smog, cancer, and even "beautiful" ladies of the evening.)

When one considers that the entire argument presented by Hamilton covered 32 pages, including detailed examples of animal behavior, one may be tempted to conclude that his thesis was no more than a "tempest in a teapot." It was obviously noticed prior to 1964 that relatives often tend to protect each other. However, the genetic factor had not been shown to be involved until Hamilton made the nexus. As a result, his work has been praised as a monumental achievement. Dawkins applauded the application of a principle that could explain sacrificial behavior that is practiced in the interest of children and other relatives. "I see the concept of inclusive fitness as the instrument of a brilliant last-ditch rescue attempt, an attempt to save the individual organism at the level at which we think about natural selection."[43] He concluded that: "His [Hamilton's] two papers of 1964 are among the most important contributions to social ethology ever written, and I have never been able to understand why they have been so neglected by ethologists."[44] Neglected?

Behavioral scientists have, in fact been greatly impressed. Gregory, Silvers, and Sutch said that "kinship theory has emerged as a major bulwark of sociobiological theory.... It can now be confidently said that altruism among kinfolk has a completely Darwinian explanation."[45] The individual apparently has a biologically based interest in the welfare of relatives. Stebbins pointed out that "sociobiologists regard nepotism as *unavoidable* (italics added)."[46] Such reactions clearly attest to a conviction that if a moral sense is to be posited, it need not extend beyond family oriented behavior. Barash, in fact, exemplifies the view of most biologists in defining kin preference among humans as an ineluctable fact:

> Kinship is a basic organizing principle in all human cultures. It is the backbone that supports *homo sapiens* society, and sociobiology provides a coherent explanation for why. We maximize our inclusive fitness when we treat relatives differently from strangers.[47]

Reciprocal altruism

Kin selection has been shown to account for a great deal of altruistic behavior. Axelrod said that "almost all clear cases of altruism...occur in contexts of high relatedness."[48] But what of assistance provided to total strangers? To charity? To an enemy? One explanation was offered by sociologists, who had made a consistent claim that "society" somehow implants in each individual all forms of knowledge (the *tabula rasa* approach). "If collective representations are prior—individuals come and go but society goes on forever—then whatever is in the individual mind must derive from the social. This would include the individual's categories of thought and perception and of course his moral values."[49]

This approach has had many adherents on the basis of its potential responsiveness to social engineering. Durkheim and others of the school of collective socialism believed that "if men were what their institutions made them, and if the very structure of their thought was determined by these institutions, then change the institutions for the better and men would likewise be perfected."[50] There is, of course, considerable appeal to such an explanation. However, the question remains: Is the source of morality the society into which one is born, or is it a type of information transported by the genetic process?

Sociobiologists offer an answer. What Hamilton proposed as the source of some altruistic behavior—kin selection—was supplemented by the work of Trivers, who argued that even assisting nonrelatives can be explained as being ultimately selfish. Here too, the impetus came from a dramatic explosion of knowledge in the fields of genetics and molecular biology, which provided solid evidence that the replication and information bearing quality of DNA strands makes these chemical machines the appropriate source of phenotypic actions.

Trivers concluded that behavior that puts the individual at risk inevitably reduces fitness. Those involved in such activity can not be expected to succeed in the competition for survival in ensuing generations. What appears to be self-sacrificial activity must be correctly interpreted as "selfish" when viewed with regard to the reproductive potential of genes. Scientific accuracy demands it!

Trivers' seminal work appeared in the 1971 *Quarterly Review of Biology*, in an article entitled: *The evolution of reciprocal altruism*. It was hailed by sociobiologists as a possible answer to their search for a scientifically credible explanation for ostensibly self-sacrificial and other supposedly altruistic behaviors. Through an extensive analysis of the behavior of many species of animal, Trivers sought to demonstrate how one could reinterpret seemingly altruistic behavior in such a way that it may equally well represent genetic selfishness, just as inclusive fitness behavior does. It could be shown to extend to friends, distant relatives, and even strangers without compromising the principle of selfishness that is the hallmark of sociobiological theory.[51]

Trivers' work provided the framework for extending sociobiological explanations to a wider range of behaviors. Dawkins was ecstatic. "There is no end to the fascinating speculation which the idea of reciprocal altruism engenders," he wrote, "when we apply it to our own species."[52] Wilson applied the theory to the activity of lower animals, saying that until Trivers' analysis appeared, while "human behavior abounds with reciprocal altruism...animal behavior seem[ed] to be almost devoid of it."[53] However, he said "armed with existing [reciprocal altruistic] theory, let us now reevaluate the reported cases of altruism among animals."[54] He did so with many forms of life, from insects to chimpanzees, including all types of relationship from social symbiosis to parabiosis, and drew the same conclusion, as had Dawkins.[55]

In the body of his work Trivers took an extreme view, contending that "reciprocally altruistic behavior can readily explain the function of human altruistic behavior,"[56] and "no concept of group advantage is necessary to explain the function of human altruistic behavior."[57] These statements obviously represent a denial of genuine altruism. He also offered an explanation for the notions of gratitude and sympathy. "Gratitude has been selected to regulate human response to [reciprocally?] altruistic acts [and] sympathy has been selected to motivate [reciprocally?] altruistic behavior as a function of the plight of the recipient."[58] Trivers also dealt with the problem of cheaters and the practice of cheating. Cheaters, he said, are simply selected against "if cheating has a later adverse effect on [their] li[ves]."[59]

He went on to explain some of the complexity involved where both blatant and subtle cheating occur. His description of such behavior, and its consequences, provides an excellent example of the extent to which sociobiologists will go in the effort to show that selfishness is the basis for all conduct. Consider this example:

> If an organism has cheated on a reciprocal relationship and this fact has been found out, or has a good chance of being found out, by the partner and if the partner responds by cutting out all future acts of aid, then the cheater will have paid dearly for his misdeed. It will be to the cheater's advantage to avoid this, and, providing that the cheater makes up for his misdeed, and does not cheat in the future, it will be to his partner's benefit to avoid this, since in cutting off future acts of aid he sacrifices the benefits of future reciprocal help. The cheater should be selected to make up for his misdeeds and to show convincing evidence that he does not plan to continue his cheating sometime in the future. In short, he should be selected to make a reparative gesture."[60]

This is an extremely complex—as well as confusing—statement, but it probably indicates that if cheaters don't reciprocate, their progeny will pay a price in reduced fitness. Draper and Harpending represent a significant number of biologists who agree. "Extreme cheaters would find few social partners and would do badly in the fitness race."[61] Thus, cheaters will be selected against.

Trivers provided a similar interpretation of guilt—an emotion that he assumed to be experienced essentially by cheaters. He proposed that "the emotion of guilt has been selected for in humans partly in order to motivate the cheater to compensate his misdeeds and to behave reciprocally in the future."[62] Thus, individuals are assumed to have an emotion which is apparently unrelated to a moral sense. He went on to speak of "sham guilt," and "mimicking sympathy and gratitude." "Sham guilt," he said, "may convince a wronged friend that one has reformed one's ways even when the cheating is about to be resumed."[63] In a discouraging (?) note he concluded, however, that "selection will favor the hypocrisy of pretending one is in dire circumstances."[64]

Trivers explained that "emotions" such as friendship "are not prerequisites for reciprocal altruism but may evolve *after* a system of mutual altruism has appeared."[65] That is, once an individual provides aid to another, a friendship may develop. The motive for providing assistance is not explained. The rationale for assisting one's enemies would seem a more difficult issue. Trivers was up to the challenge, however, saying that "one such mechanism might be the performing of altruistic acts toward strangers, or even enemies, in order to induce friendship,"[66] and "selection may also favor helping strangers or disliked individuals when they are in particularly dire circumstances."[67] Most difficult to understand is this allowance that under extreme circumstances help may be extended to those in need. It seems to suggest that there exists a moral sense that only influences behavior in extreme cases.

A summary of Trivers position, based on the many references to moral concepts, indicates that it is very similar to that of Hamilton and others who accept the kin selection hypothesis. It is that *sometimes* people have ulterior motives, *sometimes* they cheat, *sometimes* they are indignant, *sometimes* they employ sham devices, etc. Those who do not accept the sociobiological theory of reciprocal altruism could very well use Trivers' work to make their own case! But that is not the conclusion that has been drawn by Dawkins and others who are mesmerized by the sociobiological paradigm.[68]

An Evolutionarily Stable Strategy

One of the models developed for the purpose of providing support for the selfish gene hypothesis is that defined as the evolutionarily stable strategy (ESS), which is based on the principle that the fitness of a group will be enhanced if its members act as though they are aware that in the long run they will be best served by employing a cooperative strategy. Smith developed the concept, contending that "if almost all the members of a population adopt it, their (individual) fitness is greater than that of any 'mutant' individual adopting a different strategy."[69] He defined an ESS as "a phenotype such that, if almost all individuals have that phenotype, no alternative phenotype can invade the population."[70] Smith employed the "hawk-dove" game to exemplify the strategy, pointing out that for any population if individuals took an aggressive (hawkish) or submissive (dovelike) approach on every occasion their progeny would not be most apt to survive. The successful phenotype would be one that optimized the survival potential of the entire group.

Dawkins said that "if a hawk fights a hawk they go on until one of them is seriously injured or dead [while] doves merely threaten in a dignified, conventional way, never hurting anyone."[71] He concluded that over an extended period of time it would not pay individuals to act consistently as either hawks or doves. It would be most advantageous to sometimes emulate one, sometimes the other.[72] Smith, Hamilton and others contend that iterated behavior patterns that have the maximum survival value for a group will tend to become dominant—to represent a stable strategy which Dawkins described as "a pre-programmed behavior policy,"[73] pointing out that, "it is important to realize that we are not thinking of the strategy as being consciously worked out by the individual."[74]

The prisoner's dilemma

Shortly after World War II, a game known as "the prisoner's dilemma" was devised that represents an application of the ESS principle. That model, which is based on a branch of mathematics called "Game Theory," deals with the possible responses that a prisoner may make in a situation in which he may choose between blaming a crime for which he has been indicted on a partner, or remaining mute—in effect denying any involvement. The issue involves the best way to minimize the risk, or the severity, of punishment. Axelrod and Hamilton revised the game in the interest of developing a program for the purpose of improving human relations, without sacrificing the selfish gene hypothesis. They decided that some heretofore unexamined principle must operate to explain apparent altruism—perhaps some type of mutually beneficial behavior is involved. The prisoner's dilemma exercise is assumed to exemplify such a principle.

In its human form the game is played by a banker and two competing individuals. Each player holds two cards, one labeled "defect," the other "cooperate." At a signal from the banker, each player places one card face down so that his opponent does not know which card has been played. Each player may choose to "cooperate" or "defect," (i.e., refuse to cooperate). The size of payoffs and fines is arbitrary, so long as the rank order is: Most gain for defecting when one's opponent cooperates, and largest fine for cooperating when opponent defects. Thus, any player that defects (avoids cooperating) gains the largest payoff (e.g., $10), unless the other player defects also in which case each player pays a fine, (e.g., $1). If a player cooperates, but his opponent defects, the cooperator loses $5. If both cooperate, each gains $5.

In iterative situations, it has been shown in a number of research studies that individuals over time learn that cooperative behavior provides the best long term payoff. A "stable strategy" ultimately evolves. Individuals who are involved in isolated situations can be expected to take the selfish option. When the players [or any type of organism] will never meet again, the strategy of defection is the only stable strategy. However, Axelrod and Hamilton pointed out that such behavior, among humans at least, is considered socially inappropriate.[75] They conclude that the selfishness inherent in cooperative behavior represents the best long run strategy for all living creatures.

In defining the elements of the prisoner's dilemma exercise, Axelrod argued that "there is no need to assume that the players are rational.... The actions that players take are not necessarily conscious choices.... An organism does not [even] need a brain to play a game."[76] The reasons for such minimum requirements are "the evolutionary mechanisms of genetics and the survival of the fittest. An individual able to achieve a beneficial response from another is more likely to have offspring that survive and that continue the pattern of behavior which elicited beneficial responses from others."[77] In discussing human behavior, however, Axelrod said "the ability to recognize the other [person] from past interaction...is necessary to sustain cooperation."[78] He was obviously dealing here with the practice of *manipulation* or anticipated reciprocity.

Williams summarized the three sociobiological explanations of ostensibly altruistic behavior. They include kinship selection, manipulation, and reciprocal altruism. In the last case, he pointed out that, "neither the expectation nor the repayment need be conscious or even behavioral."[79] That is, reproductive fitness will be an automatic contingency of the event. He epitomized the position, saying that the behavior of a living creature is always defined:

> in relation to its single ultimate interest, the replication of its own genes. Nothing resembling the Golden Rule or other widely preached ethical principle seems to be operating in living nature....[E]volution is guided by a force that maximizes genetic selfishness.[80]

If William's analysis is correct, it would seem appropriate to recommend to young people either that they subtly "look out first for number one," manipulate others for their own interest, or practice altruistic behavior at cost to themselves or their own gene line which is evolutionarily foolish if genetic interests are paramount. Something seems amiss;

something that may be explained when the third alternative (genuine altruism) is better understood. Williams himself said that "natural selection really is bad as it seems [but]...it should be neither run from or emulated it should be combated."[81] Why does he wish to do so? To what motivational impetus can he appeal?

Rebuttal

Case after case of apparently altruistic behavior has been shown to be equally adequately interpretable as selfish. Each seems to support the argument that every behavior is ultimately self-serving. Barash said "genes selfishly looking out for themselves while appearing to be altruistic: It all seems so obvious now, but as little as fifteen years ago it wasn't."[82] Where, if anywhere, are the weaknesses in the sociobiological hypothesis?

First, there is the consistency with which almost every individual that has written definitively on the subject has *explicitly* accepted the notion of true altruism. Others have hesitated to eliminate the possibility of various types of group selection. Wilson, for example, proposed that "because the absolute number of progeny produced by surviving altruists in altruistic trait groups is greater than the absolute number of progeny produced by selfish individuals in selfish trait groups,"[83] that characteristic of behavior may survive selection pressure. Beyond this, there are a variety of serious defects in the argument.

Ambiguity

Hamilton. In his 1964a article, Hamilton made many vague or nebulous statements. He argued that individuals will *seem* (italics added) to value neighbors according to their kinship. It would be difficult to conceive of a less definitive interpretation. According to his kin selection hypothesis, the individual *must* value others in terms of their relative genetic distance. Dawkins offered an explanation for this equivocation by pointing out that "since the distinction between family and non-family is not hard and fast [the argument is only] a matter of mathematical probability."[84] This, however, does nothing to explain Hamilton's language, since in his analysis he did not restrict the recipients of altruistic behavior to those with some level of kinship. In his discussion of "giving" instincts, he referred to whomever the individual should chose to favor.

Hamilton's explanation revealed an acceptance of the existence of genuine altruism, if social behavior is any indication. He took the position that if the individual "could learn to recognize those of his neighbors who were really close relatives and could devote his beneficial actions to them alone, an advantage to inclusive fitness would at once appear."[85] He proposed that a mutation which caused such discriminatory behavior would be preserved in that gene pool. Such a mutation "itself benefits inclusive fitness and would be selected."[86] This innocent sounding explanation, like so many others provided by sociobiologists bears the seeds of its own destruction. Hamilton's claim is that if one recognizes close relatives and "devote[s] beneficial action to them" a selective mutation may occur. What does Hamilton believe causes beneficent action *before the selection of the mutation*?

Furthermore, Hamilton's claim regarding kin preference goes counter to the observations of a plethora of perceptive authors. Balzac, for example, in his book "Pere Goriot," states that there are many who:

> never do anything for their friends, or for anyone close to them simply because they owe such kindness, while in performing favors for absolute strangers they think they earn true self respect: the closer you are to them the less they like you: the more distant you are the more obliging they become."[87]

This is not by any means a rare observation. Countless commentaries by social scientists have echoed the notion. Consider such interpretations when reading Barash's comment to the effect that: "We maximize our inclusive fitness when we treat relatives differently from strangers."[88] How explain the contradiction?

Hamilton, in fact, obviously never intended to deny the existence of truly altruistic behavior. His reference to "instincts" that encourage "giving away," provides unequivocal evidence of the acceptance of some sort of a moral sense. What can one say of his contribution except that he made the point that *one* cause of *some* ostensibly altruistic behavior *may be attributed* to a desire to act in the interest of those bearing a genetic relationship? Nothing in Hamilton's thesis, nor in the research data that has been collected, makes it possible to avoid the notion that individuals are equipped with a capacity to experience positive emotions when assisting others without expectation of a reward. Those who have followed Hamilton's thesis, and have accepted his argument do not, in fact, deny that such affects occur. Hamilton's work, therefore, cannot be presumed to provide conclusive support for the selfish gene hypothesis—the contention that *all* behavior is ultimately directed solely toward the (inclusive) genetic interest of the behaver.

Trivers. An abundance of what are clearly contradictory statements were made by Trivers. In the abstract of *The Evolution of Reciprocal Altruism,* he made it abundantly clear that he accepted the existence of genuine altruism, saying quite specifically, that "each individual human is seen as possessing altruistic and cheating tendencies."[89] Later, he stated that "the human altruistic system is a sensitive and unstable one."[90] Furthermore, he referred to "*certain classes* (italics added) of behavior conveniently denoted as 'altruistic' (or 'reciprocally altruistic')."[91] However, as he began to move toward the hypothesis of reciprocal altruism he said that "*under certain conditions* (italics added) natural selection favors those altruistic behaviors because in the long run they benefit the organism performing them."[92]

At another point Trivers, made such statements as "selection will favor liking those who are themselves altruistic."[93] In this case, the term "reciprocal" was not employed. Since Trivers had equated the terms earlier (see above), suppose that his comments were modified to include that adjective. The proper interpretation would then seem to be that individuals are attracted to those who, like themselves, though they are acting in their own ultimate interests, do not reveal it. The issue was obscured further. Explanations of several heretofore confusing behaviors were offered. The first of these has been referenced by many sociobiologists, and includes some remarkable comments. Trivers said that moralistic aggression and indignation are selected for in order to:

> (a) counteract the tendency of the altruist, in the absence of any reciprocity, to continue to perform altruistic acts for his own emotional rewards: (b) educate the unreciprocating individual by frightening him with immediate harm or with the future harm of no more aid and: (c) in extreme cases, perhaps, select directly against the unreciprocating individual by injuring, killing, or exiling him.[94]

Ignoring the temptation to point out that nothing is selected "in order to" but is, rather, always only an adaptive happenstance, lest purpose be permitted to creep into the equation, in what way can the terms employed in such a statement be interpreted? The "emotional reward" must surely refer to the positive feeling that one experiences when giving or sharing. But what is the source of that reward? Unless people are repaid, Trivers said, they may become "indignant." Resentment, or hostility, is an emotion that is supposed to arise when one feels *unjustly* treated; when one feels that another person, or group, *should have* acted differently. To what, other than a moral sense, can an appeal be made? And on what grounds does one justify the practice of "educating," "frightening," or even "killing," others who don't reciprocate unless one believes that they *should*?

Trivers did allow for the experiencing of a "benevolent" emotion when assisting those in "dire need." But why should people feel motivated to assist those in exceptional need? Must they not always expect a return? Or are they energized by what Hamilton referred to as a "giving-trait"? Trivers quoted Heider, whom, he said, "analyzed lay attitudes, and finds that gratitude is greatest when the altruistic act does good."[95] Others have shown that the degree of gratitude is a function of the cost of an altruistic act to the benefactor. Does such a statement not suggest that alms recipients believe that some benefactors *should* give more?

Although Trivers dealt with the practice of cheating, he pointed out at the beginning of the article that "cheating is used throughout this paper solely for convenience to denote failure to reciprocate; no conscious intent or moral connotation is implied."[96] For those who lived through the era of the now defunct behaviorist voodooism, such a statement rings a familiar bell. It is apparently considered bad form to use the "M" (moral) word. In point of fact, Trivers said that guilt has been selected for in humans in order to motivate cheaters to compensate for their *misdeeds*. There is a strong temptation to conclude that Trivers believed that people *shouldn't cheat!*[97]

As to the notion of guilt, another curious definition is involved. Every scholarly interpretation of that concept involves either a legal infraction (not under discussion here) or "the fact of having committed an offense against moral...law,"[98] "in ethics, conduct involving a breach of moral law,"[99] or, "behavior...that is believed to be morally wrong."[100] In each case a moral element is invoked. As to "sham guilt," is the person *fraudulently* contending that he has acted in a way that he *shouldn't* have? That he has behaved immorally?

Trivers suggested that subtle cheaters, as well as those that are easily detected "should be distrusted [because they may] initiate altruistic acts out of a calculating rather than a *generous hearted disposition* (italics added), or who show either false sympathy or false gratitude."[101] Where does one begin an analysis of such a statement? Are people not distrusted because they don't do what others believe they *ought* to do? And what is "false" sympathy? It is a maxim of logic that there can be no false coins unless there are true ones. What would Trivers suggest is "true" sympathy—or gratitude? Most remarkable is his comment regarding a "generous hearted disposition." What is that, if not a conviction that one should help others without expecting a reward? And as to the hypocritical behavior that he described as characterizing those that feign extreme need, *The Random House Dictionary of the English Language*, like other such works, defines hypocrisy as "a pretense of having a virtuous character, *moral* (italics added)...beliefs."[102]

Trivers claimed that "there is ample evidence to support the notion that humans respond...more altruistically when they perceive the other as acting 'genuinely' altruistic, that is, voluntarily dispatching an altruistic act as an end in itself, without being directed toward gain."[103] He credited Heider (1958) and Leeds (1963) among others for providing instances of such behavior. Here is further evidence of an acceptance of the existence of genuine altruism. Unfortunately it conflicts with his claim mentioned above, that "the natural selection of reciprocally altruistic behavior can readily explain the function of human altruistic behavior."[104]

Trivers explanation for assisting enemies was that it may "induce friendship." It must be assumed that the recipients will either feel a sense of gratitude; an obligation to respond, (an *ought* feeling) be sufficiently foolish not to realize that they are being manipulated, or that—being human—be playing the same reciprocal game. On this basis it is assumed that friendship may develop. A tawdry relationship at best.

In the final pages of his study, Trivers once again made it clear that he accepts the notion of true altruism. He referred to the possibility that some people seem to be "genuinely altruistic," and that they operate "outside the particular reciprocal altruism system being discussed."[105] He dealt with such topics as "injustice," "unfairness," "moralistic aggression," etc., and in a depressing note he suggested that "one may be suspicious of the known tendencies toward adultery of another male or even those tendencies in one's own mate."[106] He apparently disapproves of adultery. But why? Does not such denunciation of such activity bespeak an acceptance of a moral sense?

The *piece de resistance* can be found in his expression of concern about people who take exception to the view of justice that he offered as an example of reciprocal altruistic behavior. After providing an argument to support the idea that justice operates best when each person's interests are taken into consideration, he said that those who object to his interpretation "imagine that if justice were made self-interested it would somehow become selfish."[107] His contention was that the concepts "selfish" and "self-interested" are dissimilar. If he meant only to show disdain for those who are conspicuously self-centered he was simply expressing an emotional reaction to their bad manners. If, however, he was distinguishing between the terms on the basis of a belief that one's concerns are not appropriately limited to only the biological self, and that self-interest is a relative term, he was admitting to the existence of a moral sense. There is no problem here with accepting that distinction. There is a serious one for sociobiologists.

Dawkins. It was Dawkins contention that "the fundamental unit of selection, and therefore self-interest, is not the species nor the group.... It is the gene, the unit of heredity."[108] If he was not sufficiently clear in that instance, consider others of his statements.[109] He made the claim at one point that "if you look at the way natural selection has evolved, it seems to follow that anything that has evolved by natural selection should be selfish,"[110] adding that "at the gene level, altruism must be bad and selfishness good. This follows *inexorably* (italics added) from our definitions of altruism and selfishness.... The gene is the basic unit of selfishness."[111] The principle involved is that given the fact that biologists now have a reasonable understanding of how selection works, unless genes are not the sole selection units only selfish behavior is to be expected.

In 1995, in an article essentially dismissing the role played by a superior being, Dawkins writes that "nature is not cruel, only pitilessly indifferent. This lesson is one of the hardest for humans to learn. We cannot accept that things might be neither good nor evil, neither cruel nor kind, but simply callous; indifferent to all suffering, lacking all purpose.... Genes don't care about suffering because they don't care about anything.... DNA neither knows or cares. DNA just is. And we dance to its music."[112]

Having made a clear case for the primacy of the gene and its implacable selfishness; a necessary condition of its chemical nature, he appeared in his summary, to deny all that he had written. He stated that it is possible (and that "he hopes it to be true") that man has "a capacity for genuine, disinterested, true altruism."[113] However, he said, "I am not going to argue the case one way or the other."[114] Anyone who had read his book, might be excused for coming to the conclusion that he was arguing very forcibly for "one side or the other"—the innate selfishness of all living creatures. In spite of what he had written, he pointed out that even if we "assume [?] that individual man is fundamentally selfish...we have at least the mental equipment to foster our long-term selfish interests rather than merely our short-term selfish interests."[115] This seems to be manifest evidence of his conviction that all interests are, after all, selfish. He made this absolutely clear in his statement that "pure, disinterested altruism [is] something that has no place in nature, *something that has never existed before in the whole history of the world* (italics added)."[116]

But then his astonishing caveat. "We alone on earth, can rebel against the tyranny of the selfish replicators."[117] An astonishing conclusion because it seems to represent no more than an emotional appeal. A cry in the dark. It is the sort of thing that might be said by a person who knew that a man was about to die of an incurable disease, and told him that there was always hope. From a compassionate standpoint such an approach may be admirable. However, from a sociobiological perspective it must be understood to be simply untrue. Dawkins can't categorically deny the existence of true altruism, (something that he says "has no place in nature") while claiming the capacity to "rebel" against the influence of genetic control.

On what basis does he express the hope—or the desire—for such an insurrection? Is there something beyond the control of the genes? Some *Deux ex machina*? If one's genetic inheritance proclaims "be selfish— enhance yourself" from whence arises Dawkins' feeling that an effort should be made to overcome it? Or was he saying *do it* though it should not be done? Do one's genes say, "preserve yourself—though you *should* not?" But that would represent a suggestion that individuals should sacrifice in the interest of others, and that violates sociobiological premises.

In the 1989 edition of his book, *The Selfish Gene*, Dawkins responded to complaints about what has been called his inconsistency, as expressed here; the fact that he allowed for the possibility—even the hope—that true altruism could occur. He claimed that "our brains are separate and independent enough from our genes to rebel against them."[118] One wonders where he studied biology! He argued that "it is perfectly possible to hold that genes exert a statistical influence on human behavior while at the same time believing that this influence can be modified, overridden or reversed by other influences."[119] Other influences? He is apparently willing to accept such influences as "society," or the "environment," because they can be employed without allowing for a moral sense or any such internal motivational element.

His explanation, however, was as unconvincing as his first such claim. He used as an example of the limited influence of the gene, the fact that people use contraception or other wise refrain from sex "when it is socially necessary to do so,"[120] the desire for which, he said his critics agree, "evolves by natural selection."[121] But why do people sometimes refrain? He cannot argue that they believe they *should*. They must *want* to, or they must be *willing* to do so. Consider an example such as refraining from stealing from the church poor box. Is some reciprocal action from the pastor or the congregation anticipated? Does one advertise their honesty?

And what of refraining from killing a defeated enemy. Is it assumed that the defeated individual may ultimately reciprocate? Dawkins explains. "Superficially this looks like a form of altruism."[122] However, he proposes that since people have many rivals, they may be better off if they allow any specific enemy to live because that enemy may someday fight with another of their rivals thereby benefiting [that individual] indirectly. But what of the possibility of future damage to the compassionate individual *caused by the recovered rival*?

Perhaps most revealing is Dawkins' incredible explanation for the fact that many people donate blood for potential use by others. He stated that he finds it hard to believe that reciprocal altruism could be involved because donors receive no special treatment when they, themselves, need a transfusion. "They are," he said, "not even issued with little gold stars to wear."[123] He concluded: "Maybe I am naive [!] but I find myself tempted to see it as a genuine case of *pure, disinterested, altruism* (italics added)."[124]

Pure altruism? That nonexistent sentiment? It would seem that some sociobiologists believe that blood donors, like other public-spirited people feel an urge to provide assistance to those in need. Surely there is evidence here that some believers in the fixation that no altruistic behavior is possible are naive. Especially those who have said such things as: "Scratch an 'altruist' and watch a 'hypocrite' bleed," (Ghiselen, 1974). And "'Genteel' ideas of vaguely benevolent mutual cooperation are [being] replaced by an expectation of stark, ruthless, opportunistic, mutual exploitation," (Dawkins, 1982). What shall one scratch to discover what sociobiologists *really* believe?

In 1976, Dawkins had discussed the evolution of "a gene for altruistic behavior."[125] In 1989 he apologized for what may have caused confusion, explaining that a single gene could not cause a behavior. He then pointed out that "a mutant gene in birds for brotherly altruism will certainly not be *solely* (italics added) responsible for an entirely new complicated behavior pattern."[126] How does that represent a denial of the legitimacy of altruism? Dawkins claims that his explanation "is not science fiction: it is science."[127] What sort of science is it that can do no better than provide infinitely regressive arguments?

Dawkins devoted an entire chapter of *"The Selfish Gene"* to manipulative behavior as practiced by organisms at every level of life, and his analysis is impressive. Perhaps most revealing, however, is the fact that Dawkins, and others (e.g., Axelrod), who discuss manipulation, are employing a term that carries a moral implication. Is it not considered immoral to manipulate others? Is it necessary to consult a dictionary once again to discover how a term is ordinarily employed? Most disturbing is the fact that, at least among humans, the individuals being manipulated often seem to enjoy the experience—even seeking it from time to time!

Barash. Barash proposes that society teach individuals to be altruistic. "If sociobiology is correct, we've got to be carefully taught not to hate others who are different from ourselves, because it *may be* (italics added) our biological disposition to do so."[128] His argument was that what may have been adaptive at an earlier evolutionary time "is today not only dangerous and stupid and socially reprehensible but woefully maladaptive.... We must demand that our cultural institutions, such as education and child rearing, make sure that we are 'carefully taught' to love one another. Because, sad but true, we seem unlikely to do so by ourselves."[129]

Barash's contention that people must be taught altruistic behavior was similar to Dawkins' proposal. "No human behavior," he said, "comes entirely from our genes. We must always remember that our behavior is the result of interaction between our genetic makeup and our learning, all in some specific ecological context."[130] He added: "Practically from birth we are taught that it is good to help others, and that, in fact, it is somehow reprehensible to expect recompense. If we were 'naturally' altruistic, why then all the exhortations?"[131] He completely misses the point. Nothing will be said in this text to suggest that people are—or should *always* be— altruistic. What sociobiologists deny is that people are *ever* genuinely altruistic.

Mayr, Wilson, et al. Ambiguous statements can be found in the writing of the most distinguished biologists. Mayr, who speaks of Wilson's "magnificent work, *Sociobiology: The New Synthesis,*" denies that absolute claims for genetic supremacy have been made by sociobiologists, saying "we know that almost all human traits are influenced both by inheritance and the cultural environment."[132] He asks the very question that is being raised here. "[Since] selection by its nature is a thoroughly selfish process...the great question...is how can true altruism evolve?"[133]

His explanation mirrors the claims of the most extreme sociobiologists. He first lists reciprocal altruism and kin selection which are not examples of true altruism at all. He then summarizes the section with the statement that when an individual exposes himself to danger in order to alert fellow group members to a potential threat: "This altruistic behavior benefits the survival and reproductive success of his group as a whole.... Some species are...subject to group selection."[134] Does he accept the principle of group selection? What contemporary sociobiologist would agree with such a statement?

Wilson spoke of "innate censors" and "motivators" that effect ethical premises. Consider his statement that "a genetically based act of altruism, selfishness, or spite will evolve if the average inclusive fitness of individuals within networks displaying it is greater than the inclusive fitness of individuals in otherwise comparable networks that do not display it."[135] Once again a careful reading is necessary. Is it Wilson's contention that the "altruistic" individuals must recognize their kin? Do they experience a positive affect on acting altruistically? If so, has he not indicated that positive feeling states may attend the sacrificing of one's self in the interest of another? The question here is not what the focus of such feelings may be, but whether they represent some aspect of the motivational apparatus.

Axelrod referred to the fact that kinship and reciprocity theory have been proposed in order "to account for the manifest existence of cooperation and related group behavior such as altruism...."[136] He added that "close relatedness [i.e., genetic kinship] permits true altruism.... [However he pointed out that] it can evolve when the conditions of cost, benefit, and relatedness yield net gains for the altruism-causing genes that are resident in the related individuals."[137] Thus, a form of genetic selfishness is assumed; quite the same as that proposed by Dawkins, Trivers, Hamilton and others.

At other points Axelrod appeared to support the notion of true altruism in negative statements, saying that "certain kinds of behavior that may look generous may actually take place for reasons other than altruism."[138] He also explained that his reason for believing some behavior to be based on self interest is to account for "the difficult case in which cooperation is not *completely* (italics added) based on a concern for others or on the welfare of the group as a whole."[139] Once again, in each case, the false coin presumes the existence of a genuine one. What is the source of such sentiments? How can the existence of a moral sense, and its expression in altruistic behavior be denied?

Recognition alleles

The positing of such entities as "recognition alleles" and "supergenes," and the process of "phenotype matching," are feeble efforts to provide a vehicle for the occurrence of kin recognition. Many biologists have, in fact, argued that proximity is the principle determinant of preferential treatment, which would perhaps make genetic fitness an accident of geographic factors. Consider the nature of the research that indicates that individuals are more generous to their children than to others. What point is being made? What of generosity to a wife? A lover? An in-law? A business or social partner? There is clear evidence that preference is based, at least in part, on one's familiarity with those to whom assistance is extended.[140]

In point of fact, the possibility that alleles exist that provide for the identification of relatives, as well as other humans, is accepted by few biologists. Holmes and Sherman exemplify the pessimism regarding the likelihood that they may be discovered. "An empirical search for recognition alleles would be difficult...because their existence could be inferred only after...eliminating all environmental and experiential cues, including a subject's experience with its own phenotype."[141] They added that "both Hamilton and Dawkins deemed the existence of recognition alleles unlikely due to their necessary complexity."[142]

Consider the import of suggestions that relate recognition to physical proximity rather than to genetically based awareness. The fact that genes are preferentially preserved through the procreative process may be wholly contingent. If this is the case, adopted children, since they live in conditions of regular association with their foster parents, should be more likely to be protected than natural children reared away from their natural parents. This phenomenon is considered rare by researchers, but how do they explain it when it does occur? Smith says that "from an evolutionary perspective, adoption of unrelated children seems puzzling."[143] However, he points out that "this situation has been beneficial from a *humanitarian* [italics added] point of view."[144] On what basis can he make such a statement?

Family dissidence

How shall problems that tear at the fabric of so many families be explained if kin preference is a selection factor? Burgess et al. say that "cumulative research findings indicate that the family is often a storm center marked by disaffection, disengagement, conflict, aggression, and violence."[145] A number of explanations have been offered. Burgess et al. propose that it may happen "in situations where the level of stress... exceeds the family's resources,"[146] or , as suggested by Dumas and Wahler, "when the parent has had a 'bad day'."[147] Furthermore, they argued that "differentially optimal reproductive strategies for males and females all lead to the expectation of recurring conflicts of interest in families that may, under identifiable circumstances, lead to violence...including significant maltreatment of children."[148]

Females, for example, may be badly treated "in circumstances of ecological stability such as droughts, famine, natural catastrophes, epidemics and warfare where the naturally higher mortality rates of male infants may lead to skewed sex ratios."[149] Among animals, similar evidence of intrafamily problems are reported. Trigg pointed out that "the mere recognition that the other creature is a member of the same species may be as likely to trigger hostility as a willingness to cooperate."[150]

There are any number of occurrences that seem to violate the sociobiological principle of kin preference. Along with the widely observed instances of internecine conflict is the observation that successful individuals do not propagate to the extent that financial and educational advantage would anticipate. Surely the fact that such families have a clear advantage in their opportunity to care for a relatively large number of children should result in their outreproducing their less well endowed contemporaries. Burgess et al. pointed out that sociobiologists must face the fact that "a perplexing problem for evolutionary theory is how to explain the especially low fertility of middle and upper middle [class] couples."[151] However they offer no explanation.

Friendship

Trivers suggested that generous actions—assistance to others—may lead to the establishment of friendship. As was the case with the mutation of discriminatory genes, one must ask: On what motivational impetus do individuals provide assistance to each other *before* they become friends? If no innate philanthropic sentiment can be presumed—and accepting Trivers' failure to apply the reciprocal principle—such actions must almost certainly occur by chance. Furthermore, after the initial bequest, the beneficiary must respond in the same manner—by accidentally returning the favor, since no "sense" of obligation can be imputed. This would not be sufficient to lead to friendship however, since even following a series of such adventitious engagements there is no basis on which to anticipate that a camaraderie would develop. The model requires that either the receiver of a generous act is surreptitiously planning to take further advantage of the original benefactor or that a moral sense is involved.

Consider this sequence. *A* performs an ostensibly altruistic act for *B*. Ruling out the "chance" explanation (which could never lead to the development of friendship) assume the action to represent an example of reciprocal altruism. The respondent may act in the interest of further benefit. Each is now playing a game of deceit—though no such term seems appropriate—of manipulating a relationship. And each must be foolish enough (?) to believe that in following this train of interactive experiences he will be last to profit. (In point of fact one individual would be correct according to the rules of the "prisoner's dilemma" exercise!) What friendship could be expected to arise under such circumstances?

Friendships are based on a willingness to share, to extend assistance, to sacrifice. The term must be redefined. However, the practice of redefining other-oriented behavior as always exemplifying "kin selection," "reciprocal altruism," "manipulation," etc. does nothing to solve the problem. Hull challenged such revisionism. "Before the sociobiologists are done," he said, "*aggression, altruism,* and *dominance* in ordinary English may have little in common with these terms as they are used in sociobiological theory."[152] He pointed out that "by explaining the function of certain feelings and beliefs, biologists do not 'explain them away'."[153] To contend that sharing is really a technique for receiving a reward does not eliminate the essential existence of a sense that underlies obligatory, responsible, cooperative, or other such feelings.

A phantom affect?

While biologists insist that all behavior can be shown to be selfish, in that phenotypes whose genes are most fit will have a greater representation in ensuing generations, psychologists, psychiatrists, and social workers deal with the paradoxical phenomenon of *psychopathic* or *sociopathic* behavior.[154] Although the condition is difficult to define precisely, it is ordinarily considered to describe an "antisocial or hostile individual...a person who is free from gross symptoms of a psychosis but who does not accept the mores of the larger society and acts, therefore as a troublemaker or is so defined."[155] The behavior of such individuals is assumed to represent either the lack of a moral sense (that most biologists deny exists in anyone) or perhaps limited intellectual capacity. They may not appreciate the reciprocity that seemingly altruistic behavior may generate. The psychopathic individual may lack the intelligence essential to recognizing the "sameness" in others that the majority of people do.

It is equally likely, of course, that such individuals are intellectually superior, and are simply clever enough to see that reciprocally altruistic behavior is far less efficient than taking what one wants immediately. The sociobiological response must be that the last interpretation is correct, and that society has labeled such people abnormal because of its ignorance of biological principles. Which leaves the awkward residue that psychopaths apparently lack the sense of obligation, responsibility, and orientation toward others that most individuals possess. In all instances, there is an implicit suggestion that such personalities represent a departure from the norm.[156]

Presume a normal person (one whom, according to the position taken in this text, has a moral sense) to be dealing with a psychopath; an individual who feels no sense of obligation, remorse, guilt, etc., as exemplified above. Benevolent actions would either be ignored, or be responded to with what may appear to be deference, in the hope that they would be repeated, just as is the case with a normal individual following the sociobiological thesis.

Responsive behavior would be highly unlikely, since the only explanation available to the psychopath would be that the philanthropic appearing behaviors were chance events—that no repetition of the generous acts could be anticipated except as a manipulative technique. Such responses—or lack of them—are widely observed by those who must deal with individuals adjudged psychopathic. Now consider a situation in which *both* individuals are psychopathic. What would the possibility be that a friendship may develop? None! Something is lacking in this class of individuals; something that represents a defining feature of humanity—a moral sense.

The evolutionarily stable "strategy"

There are a number of problems with the ESS principle as described by Axelrod, beginning with its label as a "strategy." It is no small matter that that term is employed. A strategy is ordinarily thought of as "a plan, method, or series of maneuvers or stratagems for obtaining a specific goal or result."[157] If anything is certain, it is that in the case of an ESS as it applies to any form of life, no *plan* could be involved. No design, aim, or intention can be ascribed. What happens is undoubtedly the result of purely contingent factors. Species, gene pools, or other classes of individuals (holons) do not develop strategies.[158] It would be clearer if sociobiologists were to replace the term *strategy* with the word *system*, when dealing with primitive forms of life, as is being done here.[159]

The significance of this issue is that several different types of influence can be shown to effect fitness. The first of these, which Dawkins and others have assumed occurs at a preconscious level, affects the most primitive animals, and even some forms of plant life. Although no speculation regarding potential long term payoff is involved, the result is that evolutionarily stable behavior patterns develop. However, such behavior is quite different from that in which deliberation is involved, the latter being the only form of behavior that is appropriately described as *manipulative*.

The more sophisticated individual behaves on the basis of desires. In order to satisfy such urges, intelligent individuals often attempt to deliberately control the behavior of others. At the level of mind, *strategy* is involved and the original ESS designation is appropriate. Profit to the individual, acquired by influencing the behavior of others is, if anything, apt to be in conflict with the ESS of that gene pool. However, at all levels of life—with or without deliberation—individuals also have a willingness to act in ways that have the effect of enhancing their gene pools or species, and they often do so at risk to themselves. That is a key element of the argument of this text, and it represents a repudiation of the sociobiologist thesis. Individuals do not act only to improve their individual fitness. The ultimate enhancement of the gene pool is the goal of altruistic behavior, which is not merely a "cooperative" venture, but involves potential sacrifice on the part of members of the pool.

Axelrod's employment of the "prisoner's dilemma," is a seriously flawed concept. It represents an example of the compulsion to demonstrate that selfishness lies at the root of all interactive behavior—even that of cooperation. It provides evidence instead, that such behavior is a threat to the human ESS. One cooperates in anticipation of a return. It pits each individual against all others. But an ESS calls for personal sacrifice when such behavior is required to serve the fitness of the group. This is a critical factor.

Cooperation between individuals of different species is apt to be of benefit to both, and thus to each species indirectly. Cooperation among members of the same species, however, which is also a class of reciprocal altruism or manipulation, may operate against the interest of a species in the long run. Although the immediate result of cooperation may be species enhancing, Smith's example of the hawk/hawk confrontation provides evidence that it is not an ESS. "When hawk meets hawk," he said, "one of them is [apt to be] seriously injured."[160] When those who cooperate only for self-enhancement persist in such behavior, such systems can be "invaded"—and have been—by individuals who respect the welfare of others.

When Axelrod spoke of an evolutionarily stable strategy among primitive organisms, he was dealing with processes that involve no decisions. Many simple creatures act solely on the basis of contingencies. "Bacteria that are normal and seemingly harmless, or even beneficial in the gut can be found contributing to sepsis in the body when the gut is perforated, implying a severe wound."[161] That is, given a situation in which the host is apt to die individuals will act in such a manner that the survival of the parasite and its progeny is enhanced. There is also no question that symbiotic behavior is evidence of pure self-interest. Here the cooperative behavior is *between* species. Each member of the interacting pair may profit, and the relevant gene pools as well. When, however, conspecifics cooperate, the "sacrifice" involved is designed to enhance the welfare of the individual who employs it at whatever cost to other members of the species.

Nowak, May, and Sigmund propose that reciprocal altruism occurs in organisms as simple as invertebrates. "If the players [e.g., *Isoptera*] occupy fixed sites, and if they only interact with close neighbors, there will be no need to recognize and remember, because the other players are fixed by the geometry."[162] But what is the motive for the first organism to act "cooperatively"? Molineux says that among microbes, for example, since female *E. coli* cells are vulnerable to a T7 plasmid, male cells (which can also be infected by the T7 plasmid) become infected and die before the phage can replicate. He considers that "an altruistic situation."[163] Shapiro says that when he fixed *E. coli* cells and sliced through them he found that cells below the dead cells seemed to be protected from harm. "My hunch," he says, "is that these cells must be protecting the cells underneath so that if something like a virus attack comes along at least they [the lower cells—not those protecting them] can survive."[164]

When Axelrod and Hamilton turned to human behavior, they were dealing with a unique situation. Humans are capable of conceptualizing situations, of predicting future potentialities, and, most importantly, of deliberately influencing the behavior of others. The problem for individuals at this level is in getting others to behave appropriately; to act cooperatively without assurance of a positive reaction. The prisoner's dilemma, thus, deals not with innate selfishness, as such, but with the capacity to reason, to recognize, and to manipulate. While such behavior undoubtedly occurs—probably more than any other—it provides no evidence that precludes the existence of truly altruistic behavior.

ESS researchers were obviously not unaware of the problem involved in determining why the first individual would risk injury or loss when the behavior options of an adversary are unknown. When speaking about cooperative behavior, they agreed that "cooperation [is] clouded by certain difficulties, particularly those concerning the initiation of cooperation from a previously asocial state and its stable maintenance once established."[165] That is precisely the question being asked here; a question that Axelrod's analysis simply does not explain.

On what basis is cooperative behavior initiated? Axelrod stated that "mutual cooperation can emerge in a world of egoists without central control by starting with a cluster of individuals who rely on reciprocity."[166] That explanation is, however, inadequate. Why should any risk behavior be initiated unless some innate evaluative mechanism which encourages self-sacrifice is involved in organisms above the level of those whose actions are purely taxic or tropistic? And in what way, if any, can human behavior be considered unique?

In the case of the game version of the "prisoner's dilemma," it is obvious that the players are conscious of the stakes. It is Axelrod's position that "the discrimination of others may be the most important of abilities because it allows one to handle interactions with many individuals without having to treat them all the same, thus making possible the rewarding of cooperation...and the punishing of defection."[167] To that extent, each player is attempting to *manipulate* his opponent. Each is making moves designed to *encourage* a certain type of response.

But that is not what Dawkins, and Trivers are referring to when they speak of reciprocal altruism among lower animals, because the individual need not know anything about the recipient of a sacrificial act. This is equally true as it refers to kin selection. Recall that Porter said that it is enough that animals act *"as though* (italics added) they were aware of their relatedness."[168] And Dawkins contended that an animal may be preprogrammed to act *"as if* (italics added) it had made a complicated calculation."[169]

In the case of a child's recognition of its mother, Porter said that the child will respond positively to any caregiver "regardless of the degree of relatedness."[170] And Smith made the point that if in any species animals habitually live with relatives "selection will favor altruistic acts that cause a net increase in the number of the relevant genes, whether or not an animal can recognize its own relatives."[171] Each of these statements refers to behavior in which individuals are not conscious, or aware of the implication of their action. Such individuals cannot answer the question of *why* they act as they do. They simply feel a compulsion to *do something,* the goal of which is genetically prescribed.

This is entirely different in principal from the "prisoner's dilemma" situation. However, the ESS does correspond to the type of response that "strategies" in simple organisms call for. In the case of humans, *decisions* are made by individuals. Among simpler organisms, the impetus to action is contingent on the adaptive value of such behavior to the individual and, ultimately to the gene pool.

There is a further complication. The prisoner's dilemma purports to provide evidence that egoists can (and do) learn to act cooperatively.

However one can discern an admission that genuine altruism exists in the tacit admission that such concepts as "trust" play a role. In trusting another to reciprocate one's safety is put at risk. That risk is greatest, Axelrod says, if other individuals are "meanies."[172] Does he really mean to assign a pejorative term to defectors? Does he believe that they are *unfair*? The acceptance of a moral sense seems unavoidable. Axelrod demonstrated the ultimate doublespeak of sociobiology when, in his 1984 text he proposed "teaching people to care about the welfare of others,"[173] and to inform them "about the fact that there is more to be gained from mutual cooperation than mutual defection."[174] The blessing of reciprocal altruism!

The evolutionarily stable "strategy," like every other explanation of selfish behavior is, in fact, blatant evidence of a departure from stability in the human species. It makes a case only for *some* selfish activity. Sociobiologists contend that self-interest is the basis for all behavior. How ironic a development, considering Smith's argument that any society that focuses on a "hawk/hawk" strategy (where each individual looks only to his own interests) is not evolutionarily stable. Unfortunately the "hawk" principle is everywhere observed in the relationships between nations, races, and creeds of every kind.

Research interpretation

In an effort to provide scientifically respectable data to support the selfish gene hypothesis, extensive studies of human as well as animal behavior have been carried out. The typical procedure in such ventures is to state a null or research hypothesis, collect data, subject the data to statistical analysis, and draw conclusions. In most studies, results are applied to groups and differences are reported in terms of degree, rather than as to whether a characteristic exists at all. Such procedures represents a comparison of gain/cost factors. Results are reported in terms of confidence levels, significance levels, etc. That process does not, however, deal with such questions as: "Is a particular hypothesis true for *every* situation?" The only possible answer to such a question would be "yes" or "no."

In order to provide evidence that is more compelling, multiple studies are done. The behavior of langur monkeys, partridges, mangebays, straw tailed willow birds, hamadryas baboons, etc., is measured on the same scale of selfishness, with the conclusion drawn that altruism is either rare or nonexistent.[175] Dawkins said that behavior practiced for the welfare of the species doesn't make evolutionary sense. Which supports the contention that altruism is a meaningless concept. Given this limitation, the critical question to be dealt with in the sociobiological analysis of research studies is:

> Hypothesis a) *"Selfish genes" account for all ostensibly altruistic behavior.*
> or
> Hypothesis b) *They account for only some such behavior.*

If research results demonstrate that Hypothesis *b* is correct, those who oppose sociobiological interpretations have made their case. What is challenged is the contention that *all* behavior is selfish. Of course, biological researchers do not accept the view that genuinely unselfish behavior exists. They are satisfied that the evidence supports Hypothesis *a*. They insist that support for that position is overwhelming. Reciprocal altruism, and kin selection, are sufficient to explain all behavior.

No one denies that some behavior is selfish—even brazenly selfish— that some is kin related, that some is manipulative, and that some is only "reciprocally" altruistic. Theological tracts recognize the existence of such behavior and excoriate much of it. However, if the selection principles that determine which phenotypes will prevail in a lineage are ultimately fixed by the activities of a coterie of mindless, chemical, robots (i.e., genes), that serve only the immediate interest of those that carry them, no genuinely altruistic behavior is possible. Sociobiologists have attempted to deal with that issue but they have not, to date, been even marginally successful. Equivocal interpretations have constantly emerged.

A paradigm shift. Until as recently as the 1960's, the general consensus among biologists as well as sociologists was that altruistic behavior is a natural feature of life. Biologists such as Wynn Edwards provided evidence in the behavior of simple organisms. Swarming and similar activities in insects was said to provide a signal to reduce, or increase the size of a colony. The logical procession was:

> *The unit of selection in all forms of life is the species.*
> *Altruistic behavior is observed in the activity of many organisms.*
> *Research data can, and should, be interpreted in such a way as to reflect that fact.*

With the discovery of the genetic code and its implications an equally powerful, but diametrically opposed hypothesis was introduced, with the same type of logical implication. Today, the belief among biologists has been altered to look upon the gene as the source of behavior, which is assumed to be totally selfish. Krebs asked:

> If we assume that the original system with which a species begins is selfish individualism...how could a system of reciprocity replace it? What would cause an individual to make an initial altruistic response; and what would induce the recipient of it to reciprocate?[176]

That is precisely the question being dealt with in this text. Why should anyone ever offer to assist another person if such an action would put the behaver at risk? However, it has not been the major concern of sociobiologists whose contention is that since:

The unit of selection in all forms of life is the gene.
Selfish behavior is observed in the activity of all organisms.
Research data can and should be interpreted in such a way as to reflect that fact.

No eclecticism is possible. If the propagation of genes is the controlling influence on behavior, the only valid conclusion must be that *no* true altruistic behavior can occur—or if it should appear as a mutation, it would not persist over time. In the case of kin selection, ostensive altruism is no more than an expression of the selfishness of the gene line. All other such behavior must be construed as representing such concepts as reciprocal altruism, manipulation, etc.

Biologists have, of course offered explanations—generally under the rubric of reciprocal altruism. However, when that could not be demonstrated beyond a reasonable doubt, they have spoken of "cultural influences," or "individual variability," etc. They have even questioned the reliability of statistical procedures that were not supportive of the sociobiological hypothesis. Wilson, for example, said, "it should be born in mind that multiple regression analysis can never prove causal relations; it can only provide clues about their existence."[177] In so doing, biologists have, in effect, made no allowance for possible falsification—a violation of a cardinal research rule. As a result, they have made proposals bordering on the bizarre.

The critical motivational flaw

When the kin selection or inclusive fitness theory is analyzed it becomes obvious that an essential element in the motivational sequence is omitted. The argument is based on an assumption regarding the influence of genes that is pure, albeit logically compelling, speculation. Dawkins said that "kin selection accounts for within-family altruism, the closer the relationship, the stronger the selection."[178] Wilson proposed that when an altruistic individual assists a brother "it will ipso facto increase the one half of the genes identical to those in the altruist by virtue of common descent."[179] These statements are almost certainly true. Altruistic behavior directed toward a relative will surely "increase half of the genes identical to those in the altruist." *But what is described in both cases is not a cause but a consequence.* The motivational sequence involved is ignored or misunderstood.

It may be that genes have a capacity to inform their "machines" to treat relatives preferentially, but *there is no objective evidence that validates that hypothesis*; only an assumption that some such process must be taking place. The logic of the argument for kin preference is, at best, circumstantial. As convincing as the evidence may appear to biologists, a step is ignored that vitiates the conclusion. The contention of sociobiologists is that:

A high positive correlation exists between the percentage of genes shared by individuals and the extent of self-sacrificial behavior practiced thus:
The cause of such behavior is the shared genes.

The problem with such an explanation is that it provides a necessary, but by no means sufficient cause for such beneficence. Shared genes do not represent the *efficient* cause of altruistic behavior among relatives. If in some situation it were the case that A was the "formal" and/or "material" cause of B, and B was necessary to the occurrence of C, (i.e., if B was the efficient cause), to assign A as the cause of C would be totally misleading.[180] Tennis players interested in causing a ball to travel in a particular direction, focus on the manner in which the ball is struck. They do not attend to its shape nor the material of which it is constructed.[181] To base a decision on such causal factors would be futile. Consider an analogy, recognizing the limitation of such a process:

A high positive correlation exists between the possession of guns and the extent to which murders are committed with those guns thus:
The cause of such murders is the guns.

The fallacy of such a proposition has been recognized by gun control opponents who have challenged it on the basis of the contention that what is not being considered is the *motive* for criminal action. A gun is necessary to the commission of a gun related crime but it is far from a sufficient cause. Considering that the existence of guns is almost certainly inevitable, while motives for such crimes are subject to educational programs, to focus on the former is highly unlikely to reduce the incidence of such felonies.

The same argument can be made for the preferential treatment of those genetically related. The basis for behavior may certainly be traced, at least in part, to one's genes. However, people do not ordinarily care for their children on the assumption that supportive behavior will increase the probability that the parent's genes will be represented in ensuing generations. They do so because of both an innate willingness to assist their children; to see their status enhanced, as well as a conviction that they *should* do so, which is based on the influence of a moral sense that encourages altruistic behavior.

The holarchic explication

In order to provide an interpretation of altruistic behavior that takes into account the proper assignment of causes, the motivational sequence—ignored by sociobiologists—must be considered. The most difficult, and often confusing, task is to appropriately identify the behavior being discussed. First, instrumental behavior in itself is insufficient to reveal an individual's goal. It must be dealt with as a cost factor, rather than as the expression of a need. Secondly, intrinsic behaviors are themselves instrumental to some outcome which is related to the enhancement of a gene pool.

When the sociobiological interpretation of kin selection is analyzed it becomes clear that the netwant/netcost aspect is unrepresented. Ingold, Crawford, Hamilton, Badcock, Charlesworth, and others all referred to *outcomes* rather than *causes* of altruism. Their (correct) contention is that the genealogical lines of individuals that assist their relatives will be enhanced. But genealogical enhancement is not the *efficient* cause of the preferential treatment of one's children. The positing of "supergenes" or "alleles" provides a possible mechanism, but does not deal with the influence of desire or evaluative mechanisms.

In the case of reciprocal altruism, the goal is obviously the welfare of the (inclusive) individual. However, such egoism accounts for only one type of behavior. When the holarchic interpretation is considered, an explanation of altruistic behavior emerges. Needs—as is the case with all desires—are shown to be prioritized. Where the desire calls for a willingness to assist others, the priority goes to those with whom one is most familiar. Proximity becomes a critical factor. The community toward which altruistic behavior is directed is, however, far more general than that proposed by sociobiologists. Kin preference is a matter of priority, rather than as the exclusive target of altruistic behavior.

There is no question that in a chronological, as well as in a rational sense, the gene is a proximate cause of the experiencing of desire. It is surely at least one cause. Furthermore, the substitution of proximity for shared genes does not solve the basic problem. Being "near at hand" is as incompetent to account for kin based altruistic behavior as is the sociobiological argument. What it does provide is an explanation for why relatives are protected or treated preferentially, not why they are treated well at all. And it offers an insight into the principle that many of the beneficent acts provided to unrelated others, as well as those that are kin related, are examples of genuine altruism. Generosity is based on a general concern for the welfare of others.

Encouraging altruism!

Perhaps the most telling arguments against the sociobiological viewpoint are the incredible recommendations that they make. Consider the usual scientific approach to a problem. First, a topic of concern is identified. An example might be the recognition of the presence of the disease known as scurvy. Studies are done that demonstrate that the lack of sufficient vitamin C is a principal cause. Physicians recommend that limes and similar types of fruit be served to those for whom such foods are unavailable for protracted periods. In the case of the problem of interpreting seemingly other-directed behavior, culturists contend that the evidence supports the claim that it is genuine. They recommend that altruistic behavior be encouraged.

Sociobiologists have concluded that "evolution is guided by a force that maximizes genetic selfishness," and "anything that has evolved by natural selection should be selfish," etc. The evidence seems clear. Recall that Dawkins said "pure disinterested altruism is something that has no place in nature, something that has never existed before in the whole history of the world."[182] Being responsible members of the scientific community, one might expect that they would recommend that the interest of the individual should take priority; that self and family preferential treatment should be encouraged.

But what have they said about the practice of altruistic behavior? *Every sociobiologist recommends that it be cultivated.* They make a persistent entreaty that people take whatever action is necessary to overcome their innate selfishness. In doing so, they manifest an incomprehensible acceptance of some rationale for self-sacrificial behavior that is apparently not genetically driven. Dawkins said that "[even if we] assume [?] that individual man is fundamentally selfish...we have at least the mental equipment to foster our long-term selfish interests rather than merely our short-term selfish interests."[183] This is palpable evidence of his conviction that all interests are, after all, selfish. Kummer proposed that moral aims may be independent of biological determinants. It may be that it is appropriate to pursue moral goals. One should simply ignore the pressure of genetic demands, and act from generous motives. But what is the basis for the existence of "generous motives"?

Recall Mayr's contention that: "The genetic component in human ethics is...of minor importance,"[184] and that people, "can adopt a second set of ethical norms supplementing, and in part replacing the... inherited norm based on inclusive fitness."[185] He proposed that "almost all human traits are influenced by both inheritance and our cultural environment."[186] Barash added that: "No human behavior comes entirely from our genes. We must always remember that our behavior is the result of interaction between our genetic makeup and our learning, all in some specific ecological context."[187] Williams, in fact, in a strong argument for struggling against genetic control, said that "natural selection...should be neither run from or emulated it should be combated."[188]

Dawkins claimed that, "our brains are separate and independent enough from our genes to rebel against them.... It is perfectly possible to hold that genes exert a statistical influence on human behavior while at the same time believing that this influence can be modified, overridden or reversed by other influences."[189] And, finally, as stated earlier: "We alone on earth, can rebel against the tyranny of the selfish replicators."[190] Each of these professionals makes the case that the desire for selfish behavior be suppressed.

But *why* should it be done? Why teach children altruistic behaviors? Either they would be violating the basic principle of selfishness, or they could employ such behavior to achieve their own ends. In fact, what "general rule" can be abstracted from such recommendations except that by appearing generous one will gain applause—ultimately perhaps an advantage over—one's neighbors? And on what basis is their reason to believe that individuals can (or *should!*) overcome their selfish nature?

The schizophrenic position in these contentions is reflected in the writings of social learning theorists who, although they agree in denying an inherent (genetic) urge to altruism, propose that such behavior be taught to children. Examples can be found in many pychological tracts. Yarrow, et al., in providing guidelines to parents said, that "generalized altruism would appear to be best learned from parents who not only try to inculcate the principles of altruism, but who also manifest altruism in everday actions."[191] In a review of recommended child rearing procedures, Grusec contended that "the more examples of altruism provided the easier it should be for children to abstract a general rule about the importance of showing concern for others."[192]

The sociobiologist argues that desires are not necessarily assuaged as soon as they arise. In many situations people are known to control inappropriate activities. Dawkins provided an example, cited above, that the fact that people use contraception or otherwise refrain from sexual activity "when it is socially necessary to do so."[193] In this instance, one desire (here obviously for sex) is in conflict with another (here apparently for social approval). The individual who refrains from sex is acting to assuage the desire for approval. Giving up the sexual experience is a cost. But, recall that every cost represents an unexpressed desire. What is the desire that appears as a cost in Dawkins' example? *What is it that an individual wants that would cause them to forego the opportunity to practice egoistic behavior?*

Sociobiologists contend that one has an expectation of reciprocity. What is sought is a return of a favor. That, in fact, must be their strongest argument. But why should individuals anticipate a compassionate response to beneficent behavior? Why should they expect others to act generously? Magnanimously? Selflessly? Is the germ of moral responsibility lying out there somewhere? The only conceivable interpretation that is consistent with the selfish gene theory, is that those who respond in kind are playing the same game; the "cooperation" game that Smith and Axelrod offered as an explanation for behavior that looks altruistic but, in fact, is based on the anticipation of some recompense. The obvious problem with such an explanation is that a clear distinction is recognized between one's emotional response when they give to others with no anticipation of a response, and when they give expecting compensation.

The entire field of sociobiology is littered with incredibly ambiguous statements, disclaimers, and blatant contradictions. The conclusion that seems most consistent with observed altruistic behavior is that other motivating forces are in operation; a desire to serve the interests of others, even when that behavior risks loss or injury, and a moral sense that evaluates propriety. In spite of every argument raised against it, however, gene selection has become an unquestioned canon. As Bertalanffy commented ruefully on considering contemporary evolutionist principles: "Like a Tibetan prayer-wheel, selection theory murmurs untiringly: Everything is useful."[194]

Summary

Heroic efforts have been made by sociobiologists to explain putatively altruistic behavior by demonstrating that it may well be no more than a subtle technique for serving selfish genetic purposes. Their interpretation of behavior is based on what appear to be sound biological principles. It is extremely difficult to challenge the contention that unless individuals behave selfishly their fitness will inevitably be reduced. They simply won't survive reproductive competition for representation in ensuing generations. The various paradigms that they have proposed however—especially kin selection and reciprocal altruism—have been thoroughly unsuccessful in proving their case. While each model undoubtedly accounts for egoistic behavior that appears to be altruistic, no one of them, nor all in concert, demonstrate that *no* behavior is truly altruistic.

Furthermore, the proponents of such explanations repeatedly make reference to a moral sense and genuine altruism. But if sociobiological theory is valid there can be no truly self-sacrificial behavior. The discovery of one example (and there are many) destroys their argument. Genic selection certainly calls for a degree of selfishness. It fits the model. Therefore, it is argued, it must be true. But only by the most extreme extrapolation can one argue that all behavior is selfish. Fortunately, for the possibility of developing a society that recognizes the role played by altruistic behavior, it is not. *The moral sense is alive and well. It exists in the mental apparatus of humans, and probably many animals as well.*

The Sociobiological Thesis

Chapter 5 Notes

1. Stebbins (1982), p. 158
2. Alper contended that "for sociobiology, no phenomenon could arise that sociobiologists would be at a loss to explain in terms of their theory." (1978, p. 198).
3. MacDonald (1988), p. 140. Turiel proposed that "among some social scientists there now appears to be a shift from environmentalism to the view that morality is biologically determined." (1980, p. 120). All behavior is assumed to be based on capacities that are generated at the genic level. As a matter of biological accuracy, that view shall find support here.
4. Wilson (1975), p. 123
5. Dawkins (1976), p. 183
6. Williams (1989), p. 182
7. Dawkins (1976), p. 215
8. Kagan quoted in Degler (1991), p. 325
9. Wade (1984), p. 265
10. Ridley (1985), p. 48
11. What is intended in this chapter is to consider how effectively sociobiologists deal with the concept of morality, and particularly with altruism. What is challenged is their claim to have discovered explanations that obviate the positing of a moral sense. Stebbins makes a point that is most instructive. "Sociobiology will flourish and increase its usefulness to the extent that it becomes an impartial quest for knowledge and a synthesis through careful, logical deductions based on all relevant facts rather than an attempt to bolster and justify certain preconceived principles or ideas," (1982, p. 411). In the case of altruism, certain "preconceived principles" may be interfering with an accurate analysis of other-directed behavior.
12. Mayr (1991), p. 182
13. *Ibid.* p. 181
14. McShea (1990), p. 80
15. Haldane quoted in Degler (1991), p. 280
16. Hamilton (1964a), p. 1
17. *Ibid.* p. 10
18. *Ibid.* p. 13
19. *Ibid.* p. 16
20. ———. (1964b), p. 19
21. Smith (1989b), p. 37
22. *Ibid.* p. 38. He argued that "populations do not have fitness...because populations do not reproduce and hence there is no 'generation' on which

fitness could be measured," (*Ibid.*). He thus tied fitness to collections of individuals, without reference either to their genes, their population or their species membership.

23. Li & Grauer (1991), p. 237
24. Hamilton (1964a), p. 20. Several examples of the application of the theory were provided. Because of the small amount of effort and the considerable gain involved in mutual grooming among animals, he said that, "the degree of relationship...need not be very high before the condition of inclusive fitness is fulfilled; and for grooming within some families of monkeys, it is quite obviously fulfilled," (*Ibid.*, pp. 20-21). Biologists, in general, agree with Hamilton. Among nonhuman primates, kin selection is assumed to be clearly manifested. Wilson, from his study of social insects, argued that "the primary 'goal' of a social vertebrate is the best arrangement it can make for itself and its closest kin within the society,". (1975, p. 381).
25. Degler (1991), p. 281. Two points must be considered here. First, Degler's argument deals only with the hypothesis that some—*not necessarily all*—altruistic behavior can be explained on the basis of the inclusive fitness principle. More importantly, if reasonableness is to be applied, how would such defenders answer the equally mystifying question: If there is no moral sense, why do people so often act philanthropically where no kinship or potential reward can be discerned? Unless such "selfish" theories explain *all* ostensibly altruistic behavior, that problem must be dealt with. Hamilton did not address the issue.
26. Ortner (1983), p. 134
27. Badcock (1991), p. 64
28. Charlesworth (1988), p. 44
29. Ingold (1986), pp. 282-283
30. Hamilton (1964a), p. 25
31. Holmes & Sherman (1983), p. 55
32. Smith (1988), p. 279
33. Segal (1988), p. 175. Rushton et al. contend that "organisms are able to detect other genetically similar organisms...and to exhibit favoritism and protective behavior toward these strangers," (1984, p. 179). Segal added that "naturalistic observation of interactions within nonhumans social primate groups has revealed that the frequency of spatial proximity and social exchange (including altruistic acts) are associated with degree of kinship," (1988, p. 83).

She also pointed out that among relatively primitive humans the same sort of behavior has been observed by anthropologists. "These studies are provocative, because they direct attention to a generally neglected variable (genetic relatedness) as possibly underlying the apparent 'social specificity' of peer related activities," (*Ibid.*).

34. Barash (1979), pp. 152-153
35. Badcock (1991), p. 104
36. Porter, et al. quoted in Smith (1988), p. 281
37. *Ibid.* p. 280. In still another, when subjects were asked to match photographs of newborns with those of several recent mothers, including the baby's real mother, "Subjects performed at a better than chance level...indicating that there is a degree of actual similarity in the facial features of newborn humans and close kin, thus providing potential cues for phenotype matching [mentioned above] to occur," (*Ibid.*).

Other evidence of visual recognition include data showing that infants less than 4 days old prefer to look at their mother's face, and that "mothers can accurately identify their own infants from amongst an array of color photographs of neonates all of the same sex and less than 33 hours old," (Porter, 1987, p. 184). Additional data from a series of researchers is provided. "Neonates will work (suck on a nutritive nipple) to produce the voice of their mother in preference to that of another woman," (DeCasper & Fifer referenced in Porter, 1987, p. 185). "Mothers, in turn, recognize their own infant's cries as early as the third day after delivery," (*Ibid.*). And "as the infant begins to show evidence of recognizing its mother...mothers report more positive feelings toward their baby," (*Ibid.* p. 178). However, in the case of infant research findings a caveat is mentioned. "Any advantages that accrue to the infant from recognizing and reacting preferentially to its caregiver should be similar regardless of their degree of relatedness to one another," (*Ibid.*) which raises again the specter of proximity as an alternative explanation.

38. Segal (1988), p. 189
39. *Ibid.*
40. *Ibid.* p. 186
41. *Ibid.*
42. Williams (1989), p. 192
43. Dawkins (1982), p. 187
44. ———. (1976), p. 97
45. Gregory, Silvers & Sutch, (1978), p. 11
46. Stebbins (1982), p. 393
47. Barash (1979), p. 141
48. Axelrod (1984), p. 88
49. Fox, R. (1989), p. 64

50. Durkheim referenced in Fox, R. (1989), p. 67
51. Trivers accepted the caveat that "there is no direct evidence regarding the degree of reciprocal altruism practiced during human evolution, nor its genetic basis today.... However," he said," it is reasonable to assume that it has been an important factor," (1971, p. 48). A "reasonable assumption" is the basis for the theory that he proposed!
52. Dawkins (1976), p. 202
53. Wilson (1975), p. 120
54. *Ibid.* p. 121
55. As life evolves from the more primitive to the more complex, he said, "the key properties of social existence, including cohesiveness, altruism, and cooperativeness, decline," (Wilson, 1975, p. 379). The result has been that through an increase in intelligence, humans can "profitably engage in acts of reciprocal altruism that can be spaced over long periods of time, indeed over generations," (*Ibid.* p. 80). For Wilson, the power of Trivers' hypothesis makes it possible to explain even the most extreme cases of altruism as ultimately designed to profit the individual or its progeny.
56. Trivers (1971), p. 48
57. *Ibid.*
58. *Ibid.*
59. *Ibid.*
60. *Ibid.* p. 50
61. Draper & Harpending (1988), p. 365
62. Trivers (1971), p. 50
63. *Ibid.*
64. *Ibid.*
65. *Ibid.* p. 49
66. *Ibid.* p. 52
67. *Ibid.*
68. Consider a few examples. Masters wrote that "sociobiologists account for cooperative behavior—and especially for altruism—in terms of the long range or rational advantage of individual participants," (1981, p. 138). Wilson contended that "if a trait is to be selected it must increase the fitness of the bearer relative to the fitness of other members of the population.... Conversely a trait that decreases personal fitness may be selected if it does even more damage to other members of the population," (1980, p. 5). And recall Dawkins' claim. He certainly did not equivocate! "A minor revolution has taken place.... 'Genteel' ideas of vaguely benevolent mutual cooperation are replaced by an expectation of stark, ruthless, opportunistic, mutual exploitation," (1982, p. 55).

69. Smith (1980), p. 25
70. Smith (1989a), p. 126. Axelrod added that a strategy is evolutionarily stable if the genetic structure of the individuals that employ it "cannot be invaded by a rare mutant adopting a different strategy," (1984, p. 11).
71. Dawkins (1989), p. 70
72. He made the further point that: "In the hawk-dove game, it is supposed that the fitness of an individual is determined by one or more pairwise [hawk/dove, dove/hawk] interactions with random partners," (*Ibid.* p. 132). Casti developed mathematical models of the ESS process, pointing out that "the evolutionary trend of [a] population is defined in terms of phenotypic *behaviors*, implicitly assuming that these behaviors are observable indications of underlying genetic variations," (1992, p. 9).
73. Dawkins (1976), p. 74
74. *Ibid.* p. 132. As in Maynard Smith's analysis, Dawkins (1976) said, "the word 'strategy' refers to a blind unconscious behavior program," (*Ibid.* p. 162). He did, however, use the word "policy" which is equally purpose laden. And, more importantly, though he continually denied the requirement that any sort of deliberation be involved, many of his examples included such terms as "cheating," "retaliation," etc.
75. Axelrod & Hamilton exemplified a response to a player that always defected (failed to cooperate), by stating that "in everyday life, we may ask ourselves how many times we will invite acquaintances for dinner if they never invite us over in return," (1989, p. 122). That is, if they never reciprocate. Although the development of evolutionarily stable strategies is believed to occur at even the simplest levels of life, among humans it is assumed that reciprocal altruism, or "manipulation" is commonly involved. The issue of manipulation has been dealt with by a variety of theoreticians. Williams said that "anyone who makes an anonymous donation of money or blood or other resources as a result of some public appeal is biologically as much a victim of manipulation as the snapper in the jaws of the anglefish," (1989, p. 193). "Manipulation is [one] reason why one individual may provide benefits for another. [One] interesting kind of manipulation results from deception," (*Ibid.*,) p. 192. The fact is that Axelrod's "cooperative" behavior is not a form of altruism at all.
76. Axelrod (1984), p. 18
77. *Ibid.* p. 22
78. *Ibid.* p. 139
79. Williams (1989), p. 193
80. *Ibid.*
81. *Ibid.* p. 196

82. Barash (1979), p. 133
83. Wilson (1980), p. 520
84. *Ibid.* p. 101
85. Hamilton (1964b), p. 21
86. *Ibid.* p. 22
87. Balzac (1994) p. 27
88. Barash (1979), p. 141
89. Trivers (1971), p. 35
90. *Ibid.* p. 48
91. *Ibid.* p. 35
92. *Ibid.*
93. *Ibid.*
94. *Ibid.*
95. Heider quoted in Trivers (1971), p. 49
96. Trivers (1971), p. 36
97. The 1987 *Random House Dictionary of the English Language* defines cheating as defrauding, swindling, or deceiving. It is hard to believe that Trivers could support a definition that disagreed with such frankly morally inappropriate behaviors.
98. *The Random House Dictionary of the English Language* (1987), p. 849
99. Runes (1962), p. 120
100. Wonderly (1991), p. 273
101. Trivers (1971), p. 51
102. *The Random House Dictionary of the English Language* (1987), p. 943
103. Trivers (1971), p. 51
104. *Ibid.* p. 48
105. *Ibid.* p. 53
106. *Ibid.* p. 54
107. *Ibid.* p. 38
108. Dawkins (1976), p. 12
109. The selections referenced here are taken from Dawkins' 1976 text, *The Selfish Gene*. He has written extensively since then, particularly in 1982 with *The Extended Phenotype*, in which he attempted to further explain his position, and in a November 1995 article "God's utility function" in the *Scientific American* journal. However, the principles have not been altered. This section has not been written to attack Dawkins' work, but only to demonstrate the kinds of problem that writing of that type has created.
110. Dawkins (1976), p. 4
111. *Ibid.* pp. 38-39

112. Dawkins (1995), pp. 80-85
113. *Ibid.* p. 215
114. *Ibid.*
115. *Ibid.*
116. *Ibid.*
117. *Ibid.*
118. _____. (1989), p. 332
119. *Ibid.*
120. *Ibid.*
121. *Ibid.*
122. *Ibid.* p. 68
123. *Ibid.* p. 230
124. *Ibid.*
125. _____. (1976), p. 64
126. _____. (1989), p. 281
127. *Ibid.* p. v
128. Barash (1979), p. 154
129. *Ibid.* This is an excellent example of the absurd argument that since we cannot love each other, let us demand that others (e. g., teachers) encourage loving behavior. Are such individuals from another planet? Do they perhaps possess a moral sense? A propensity for altruism?
130. *Ibid*, p. 159
131. *Ibid.* p. 167
132. Mayr (1991), p. 155
133. *Ibid.*
134. *Ibid.* p. 157
135. Wilson (1975), p. 118
136. Axelrod (1984), p. 89
137. *Ibid.* p. 96
138. *Ibid.* p. 135
139. *Ibid.* p. 6
140. It could certainly be argued, *on the same principle*, that altruistic behavior toward one's relatives is as selfish as that directed toward strangers. After all, I have twice as many of my own genes as do any of my children, and by receiving assistance from them in return for my "generosity" I may be able to father more carriers of my genes—undiluted by those of a partner of happenstance!
141. Holmes & Sherman (1983), p. 51
142. *Ibid.*
143. Smith (1988), p. 281

144. *Ibid.*
145. Burgess et al. (1988), p. 293
146. *Ibid.* p. 308
147. Dumas & Wahler referenced in Burgess et al. (1988), p. 310
148. Burgess et al. (1988), p. 311
149. *Ibid.* p. 312
150. Trigg (1983), p. 141
151. *Ibid.*
152. Hull (1989), p. 278
153. *Ibid.* p. 258
154. Hare (1970) suggested that the term psychopathy, though still widely employed was replaced by sociopathy because it (psychopathy) was so "cumbersome."
155. Harriman (1974), pp. 154-155
156. In 1980, Dr. Kupfersmid, then employed at a children's psychiatric hospital in Cleveland, Ohio, and I explored the emotional states of individuals that had been labeled "psychopathic." We were reacting to a number of diagnostic reports that stated that "anger" had been displayed by some of the residents. From our theoretical perspective, anger can no more be experienced by a psychopathic person than guilt, since each emotion involves the conviction that some person, or persons *should*, or *should not* have performed or be performing some action. We distinguish anger which we define as being based on "the belief that some morally responsible agent is behaving in a way that illegitimately frustrates a desire," (Wonderly 1991, p. 274), from frustration *per se*, which refers to interference without a moral element, as when one cannot open a locked door, or find a misplaced shoe—where no responsibility is assigned.

    We discovered that in every instance that the term *anger* was employed, the child was, in fact, experiencing only *frustration*, but had *learned* to employ the term anger. It is our hypothesis that most individuals find behavior that includes a respect for the welfare of others to be natural and desirable and that a feeling of anger must be associated with some moral propensity. Such an assumption is obviously inconsistent with the laws of human behavior proposed by sociobiological theorists.
157. *The Random House Dictionary of the English Language* (1987), p. 1880
158. Smith explained that when the members of a population have a variety of phenotypes, "these phenotypes are often referred to as *strategies*," (1989b, p. 126). However, the term has an unfortunately misleading connotation. The ESS of primitive organisms is not based on any goal whatsoever. It is a fortuitous occurrence.

159. The systems concept was developed in Chapter 4, where it was pointed out that only in the case of organic entities is directionality involved. Individuals are related to their gene pools subordinately. However, the actions of individuals in meeting the evolutionary demands of gene pools, represent blind responses to differentially rewarding circumstances in many cases.
160. Smith (1986), p. 76
161. Axelrod (1984), p. 103
162. Nowak, May & Sigmund, (1995), p. 80
163. Molineux, (1995), p. 103
164. Shapiro (1995), p. 102. Oliwenstein makes the critical point that "[although] sacrifice and altruism are concepts not commonly associated with bacteria, nor are bacteria known for going willingly to their death...researchers are now discovering that everywhere you look in the bacterial world, rather than immortal life you find death.... Whether you call it suicide or altruism programmed cell death: appears to violate evolutionary good sense," (1995, pp. 99-100).
165. Axelrod & Hamilton (1989), p. 138
166. Axelrod (1984), p. 69
167. *Ibid.* pp. 94-95
168. Porter (1987), p. 176
169. Dawkins (1976), p. 101
170. Porter (1987), p. 178
171. Smith (1986), p. 58. Dawkins, in fact, referred to a wide variety of behaviors that are carried out without awareness of their cause by the strategists that employ them. "We are," he said, "entirely familiar with the idea of unconscious strategists, or at least of strategists whose consciousness...is irrelevant," (1989, p. 228).
172. Axelrod (1984), p. 63
173. *Ibid.* p. 141
174. *Ibid.*
175. Problems with procedures such as the combining of data where it may not be appropriate, and ignoring research results that disagree with a particular research hypothesis are discussed in the text, *An author's guide to publishing better articles in better journals in the behavioral sciences.* Kupfersmid & Wonderly (1994).
176. Krebs (1987), p. 86
177. Wilson (1975), p. 524
178. Dawkins (1976), p. 101
179. Wilson (1975), p. 118

180. In Aristotelian logic, causes of an entity, a situation, or an event, were sought in four areas, "the material cause (that of which a thing is made), the efficient cause (that by which it comes into being), the formal cause (its essence or nature, i.e. that which it is) and the final cause (its end, or that for which it exists." (Runes 1962, p. 21)
181. The fallacy in such an explanation has been lamented by many philosophers. Hospers pointed out that "We do not count it an explanation if it is said that the hen sits on her eggs *in order to* hatch chicks, because we have no indication that the hen does so with this purpose in mind; even if this is true, we do not know it, and therefore we cannot use it as a law in our explanation," (1966, p. 189). That is precisely the argument that has been advanced in this text for denying the claim that people assist their children, and other kin, because they share a portion of their genes!
182. Dawkins (1976), p. 39
183. *Ibid.*
184. Mayr (1982), p. 82
185. *Ibid.* p. 83
186. Mayr (1991), p. 155
187. Barash (1979), p. 159
188. Williams (1989), p. 196
189. Dawkins (1989), p. 332
190. *Ibid.* p. 215
191. Yarrow, Scott & Waxler, (1973), p. 226
192. Grusec (1981), p. 74
193. Dawkins (1989), p. 332
194. Bertalanffy (1952), p. 95

Chapter 6

The "Selfish" Gene Pool

The best basis for a scientific judgment of whether a specific act is good or bad is to consider its effect on all mankind.

Lee Dice

Species and Gene Pools

From the time of Greek biological studies—and earlier—the "reality" of such entities as species, populations, genera, etc., has been disputed. Realists accept their legitimacy. Nominalists do not. Nevertheless individuals have been identified as belonging to *something* in the work of many eminent biologists. Although the literature focuses on species and populations, if for "species" one reads "gene pool" the relevant issues should be discerned.

Dobzhansky pointed out that "typological thinking (the study of classes of entities such as species and gene pools) has become habitual not only among biologists and other scientists but among the general public as well."[1] Rosenberg, however, contends that species definitions are so broad as to make any analysis ambiguous.

> There is at present no general agreement on an explicit definition of the term "species".... There are: The so-called biological species notion

which hinges on relationships among conspecifics; the evolutionary species concept, which stresses the unitary response of a lineage to selection; and the ecological treatment of species, which identifies them indirectly through the specification of their position in an ecology.[2]

Mayr preferred the first definition. "Anyone working with living populations restricted to one place and one time, finds any species concept other than a biological one to be unsatisfactory.... The species is not an invention of taxonomists or philosophers, but...has reality in nature."[3] His thesis was that those who object to the notion of a species as an existent think of species across time—where they do change—but at any fixed time and place, he argued, they are real.[4] A species shall be defined here simply as an entity which is comprised of a number of isolated gene pools. It is more general than a gene pool in its inclusion of many populations of similar, but nonidentical, genotypic components.

The difficulty associated with the species concept is, of course, not overcome by the substitution of such terms as *population* (a group of organisms living in a specific region) *deme*, (an aggregate of interbreeding individuals), or *gene pool*. In every case, the nominalist objection can be raised. By failing to find a common ground biologists have left themselves vulnerable to criticism. However, when the subservient role of individuals is considered it should be clear that the *precise* demarcation of the source of genetic influence is not essential.

It is simply inaccurate to reference a species, a population, or a gene pool as a "starting point" in the progression of life. Gene pools and species—like individuals—are "born" live for some period of time, and ultimately die. They, too, represent no more than an intermediate level in the chain of life. Perhaps the trepidation of those who fear the acceptance of a gene pool as "real" can be assuaged by the concession that *no deliberation is assumed to be involved* in the influence that such an entity has on the behavior of its constituent phenotypes.

The fact that benevolent behavior is evaluated on a moral scale as representing the practice of serving the needs of others does not necessarily pinpoint the location of those "others" in time or space although the gene pool may be understood to be the immediate biological progenitor of each gene complex. The fuzziness lies in the fact that a repeated admixture of DNA across eons has resulted in the creation of an untraceable lineage. It is sufficient to appreciate that a moral sense exists and that it is based in part on genetic factors. Furthermore, that sense is the evaluative mechanism that influences altruistic behavior.

The "Selfish" Gene Pool

As a consequence of the evolution of such interactive and self deprecati-ng behavior, patterns have emerged that have stabilized gene pools, populations, and species in *evolutionarily stable systems* in many forms of life. The holonic/holarchic nature of living existence requires that parts and instances (e.g., phenotypes) serve the needs of the systems that they manifest. It was pointed out earlier that they do so through the perception of, and reaction to stimuli that have been described acting as "signal" systems (i.e., desires—both self-serving or egoistic, and self-deferential or altruistic).[5] In higher animals, and especially in humans, the individual evaluates potential behavior in terms of the outcome that may be generated. Several principles that provide arguments for the existence of genuine altruism and the priority of the gene pool shall be developed in this section.

- Every argument mounted by sociobiologists can be countered with an interpretation employing the same logic, that legitimizes genuinely altruistic gene pool enhancing behavior. (The data provided by sociobiologists such as Dawkins, Hamilton, and Trivers demonstrates this.)
- All behavior, including that carried out in the interest of others, has evolved on the basis of the *function* it serves, which is not always individual fitness but often serves the interest of some more comprehensive form of life. Every aspect of a behavioral episode is, thus, reducible to a status condition and ultimately to the potential enhancement of a gene pool. *The gene pool is, in fact, the "selfish entity"*.
- Desire, capacity, and opportunity as well as associated costs must all be involved when a behavior occurs. The individual must be willing to pay some price to get what is desired. That price often includes instrumental behavior. *All behavior involves cost.*
- Neither capacity, opportunity, or need alone—nor in concert—can create a desire. Although the individual must believe that both the capacity and the opportunity to act are present if a behavior is contemplated, sacrificial behavior, (like all others) cannot be explained as a consequence of behavior that ignores the contribution of desire.
- Evidence for true altruism can be found in the extensive data regarding the practice of unselfish behavior among all classes of people. In such instances, the self-transcendent urge and the moral sense make a joint contribution.

The Evidence

It was pointed out in Chapter 4 that all behaviors are "outcome oriented." They are motivated by desires that are associated with the achievement of status conditions. Warnock says that "A person is moral, by and large, exactly in proportion as he really wants to be."[6] Given that state of affairs, when is it appropriate to consider behavior altruistic? Should a distinction be drawn between behavior carried out when the consequences of the action are taken into account by the behaver and that in which they are not? Should the monkey that barks at the approach of a predator be considered altruistic if the animal is unaware of what may eventuate? What of the charitable contribution made without concern as to how it is spent? Is that behavior less worthy than a gift where the donor specifies its use?

There is, of course, considerable difference of opinion regarding the possible distinction between human munificence and the professedly altruistic behavior of simpler animals. In dealing with animal behavior, self sacrifice- is often specifically referenced. Wilson, for example, said that "social insects contain many striking examples of altruistic behavior,"[7] referring particularly to what appears to be suicidal behavior. Such behavior "seems to be a device specifically adapted to repel human beings and other vertebrates."[8] The insects to which Wilson referred are certainly not aware of the purport of such action.

Rachels reports on a series of experimental studies in which it was demonstrated that "a majority of rhesus monkeys will consistently suffer hunger rather than secure food at the expense of electroshock to a conspecific."[9] He concluded that: "Taken together, these results provide impressive support for the view that rhesus monkeys are altruistic.... [It shows]that their behavior is influenced by factors similar to those that shape altruistic behavior in humans."[10] Those "factors" include a desire, that they surely do not understand, and, most significantly, a willingness to make personal sacrifices in certain situations and perhaps a sense that such behavior is appropriate.

Ritualized behaviors (which seem analogous to social conventions among humans) are observed in the activities of many animals, especially primates, though most primatologists insist that animals are incapable of what is ordinarily defined as moral conduct. Kummer, commenting on the results of research on animal behavior such as that described above, said that since clear evidence of animal morality has not been discovered, "the possibility remains that monkeys and apes are indeed amoral."[11] There is undeniable evidence that animals manifest a great assortment of socially prescribed behaviors. Students of animal behavior, however, refer to most such activities as moral "analogues"—not being truly moral—though they look very much like it.[12]

Among the lowest animal forms, allegiance to one's offspring is only rarely discerned (and then usually briefly). It makes its appearance where the care and feeding of the young requires the continuing effort of a parent. Even here, loyalty is usually reserved for members of the same species, symbiotic relationships representing only the mutual advantages that such relationships provide. A man who destroyed a human to save a tree would generally be considered immoral. By contrast, many would accept as laudable the behavior of one who destroyed other individuals to save his family. Dawkins, in fact said: "Genes are correctly understood as being selected for *their capacity to cooperate*."[13] The problem is to determine whether this latter form of behavior is perverse or part of the natural order.[14]

Societies and cultures of all kinds, have based laws, customs, and interpretations of ethical behavior on a moral sense which makes possible the evaluation of behaviors on a scale of propriety; on their " rightness or decency." In every instance it can be demonstrated that such behaviors are altruistic in nature. They represent a commitment to the serving of the interests of others—the family, the community, the State—before those of the self. The extent to which such behaviors occur has been investigated by a large number of researchers.

In the case of humans, it is suggested that the ability to deliberate, and to draw conclusions about the consequence of behaviors has resulted in the emergence of genuine altruism. Mayr believes the issue of self- sacrifice by animals to be critical to an understanding of human morality, arguing that "*the shift from an instinctive altruism based in inclusive fitness to an ethics based on decision making was perhaps the most important step in humanization.*"[15] And Warnock proposed that "what makes 'people' eligible for consideration, and sometimes for judgment, as moral agents is that they are in a certain sense rational and not that they constitute a particular biological species, that of humans."[16]

Ehrlich and Holm contend that the chemistry of evolutionary development has consistently been modified by the impact of language, communication, and other particularly human attributes. "Indeed, from the very beginning of culture, man's evolution has been characterized by the interactions of biological and cultural evolution."[17] The common denominator for biologists, is potential damage to the altruist in the case of animal behavior, but in the case of humans the concept is modified to reference any action that serves another's interest, where there is some—but not necessarily calamitous—sacrifice. The reference is to a more general concern for the welfare of others.[18]

Guisinger and Blatt report that "there is a powerful instinctive drive to aid others in distress, a drive that can be detected even in newborn infants."[19] Researchers have found that "day old babies become distressed when they hear another baby crying, [and they have found] high levels of helping behavior in children between nine months and two years of age."[20] They conclude that: "Children show the emergence of helping behavior among the first activities of life."[21] Other researchers have reported that at around 18 months of age, in an "explosion of prosocial behavior," children have been observed to "help, share, protect, defend, comfort, console, give simple advice, and mediate fights [in] an apparent attempt to set things right for the victim."[22] It would be presumptuous to contend that such behavior is carried out in the expectation of some reward. Children are simply responding to a feeling of discomfort; to a negative signal. They are surely incapable of appreciating the possible consequences of their actions.

By contrast, much altruistic behavior involves some knowledge of potential outcomes. Professional policemen, firefighters, and servicemen and women put themselves in harms way on a regular basis, confident that their actions are of value to the community. Individuals respond generously to appeals that include information about the way in which their contributions will be used. Hundreds of thousands—perhaps millions—of people die annually in the struggle to defend a country, a religion, or other shared characteristic. In many cases, positive emotions are experienced prior to, during, and after such self-sacrificial behavior. It takes little insight to recognize that some (though obviously not all) such behavior may be construed as other than self or kin oriented. The *willingness* to sacrifice indicates the existence of a desire to behave in ways that will enhance the status of others—including those not related by blood.

When one realizes that the function of all behaviors is to enhance a gene pool, it becomes clear that the only difference between altruistic behavior based on a consideration of consequences, and that where the donor is unaware of the possible results of their benificence is that humans have the capacity to appreciate the significance of their behavior. The process is no different in principle than that of the role played by the bear that takes on fat "for the coming winter" or the squirrel that hoards nuts "to avoid a possible shortage of food in the months ahead." The individual cannot appreciate the source of the desire; only that a desire is experienced. In every case the function of such behavior is to enhance the relevant gene pool through behavior that enhances the individual. The critical factor is the willingness to risk or sacrifice. The latter type of behavior when call upon exemplifies the practice of altruism.

Extensive documentation is available to support the contention that gene pools have priority in the selection of behaviors. Looking closely at examples of conventional social behaviors one discovers that in every instance a respect for the interests of others is manifested. Such practices often place the behaver in a position of self-induced subordination. Most importantly, in a well integrated society no one performs in such a way *only* because of the assurance of a *quid pro quo*. There is no resentment felt at being the last person to act graciously. The implication of all social regulations is that under some conditions individuals performing social—even ritualistic—acts would behave in the interest of the individual or institution toward which the action was addressed, though such an action was at cost to themselves.

Of course rules of conduct are *taught*, and it is commonly stated that altruistic behavior will ultimately provide the "best arrangement for all." Arguments of this type are based on the contention that it is reasonable to be polite, which is quite a different proposition than the notion that it is moral to act "appropriately." Such an approach works—especially with children—(be nice to Robert and he will be nice to you) because the fallacy of the proposition is not recognized. Would one teach a child that if Robert is not nice, that child should be disagreeable to Robert—which would be a rational consideration?

A distinction is often made between *courteous* or *mannerly* and altruistic behavior. The latter is understood to refer to action taken that puts the interest of others before that of the behaver. A risk of loss, including injury or death may be involved. Courteous or mannerly behavior require submission to no such peril, so long as no threatening situation is involved. However, the source of each type of behavior lies in an awareness, and respect for the value of life, imparted by the genetic process; an implicit obligation to act in the interest of whom or whatever the courteous behavior is directed.

Consider the donation of part of one's treasure to the less fortunate. The civility associated with breaking bread, for example, and thus taking only that which one is apt to eat has a long history which is based on altruistic principles. Aresty quotes from a 14th century book on manners (the *Boke of Curtasye*).

> Bite not thy bread and lay it down
> This is not curtsey to use in town
> But break as much as you will eat
> The remnant to the poor you shall lete (leave).[23]

Her contention is that "what was good manners in medieval times and even earlier still prevails in polite behavior today, if it is based on a moral premise."[24] The only concern here is with her disclaimer, since *all* social graces are founded on moral principles. Aresty, for example said that "to show respect for age has its roots in ethics."[25] However, she claimed that behavioral forms that represent types of etiquette are extremely flexible: "A rule today, outmoded tomorrow."[26]

Ford took the same position. "Although new models of behavior evolve from trends into tradition [and] others become obsolete...the underlying core of etiquette remains graciousness and consideration for others."[27] However some practices persist over long periods of time. "The rule of breaking bread into morsels before eating has never wavered"[28] said Ford although the maintenance of gracious behavior is difficult because "man is not by nature considerate."[29] This is, of course, the sociobiologists claim—one that Ford's reference to the "code of etiquette" would not support.

Altruism and the law

As with other forms of social interaction, legal practices are based on moral premises and altruistic presumptions. From the code of Hammurabi, through the *ius* of Rome, to modern judicial procedures, civil as well as criminal statutes have been based on such concepts. "Jurisprudence—wisdom in the law—was defined in the *Digest* of Justinian (A.D. 533) as both a science and an art: the 'science of the just and the unjust,' and the 'art...of the good and the equitable.[30] Every descriptor represents an example of a moral concept as expressed in an ethical dictum. Although civilization has brought about an expansion and modification of rules of behavior, the basis for judicial regulations has remained the same—the settling of litigious issues on morally derived principles, and the conviction that it is appropriate for one litigant to suffer in the interest of others.

Political philosophers, from Plato to Baird, have argued that the State, or highest civil institution to which an individual ordinarily feels an allegiance, plays a critical role in the determination of the moral fiber of its citizens. It possesses life and death control over behavior, institutes rules and codes, and defines sanctions associated with violations. Plato considered that "the purpose of its [the State's] constitution and its laws is to bring about conditions which will enable as many men as possible to become good."[31] Aristotle agreed, proposing that the State, like any community, exists for a specific end. "In the case of the State, the end is the supreme good of man, his moral and intellectual life."[32]

In spite of reservations regarding the limit of political responsibility, there has been almost universal acceptance of the principle that political leaders should provide moral leadership. Locke proposed that government is a "moral trust," and regarding the particular case of the political ideology of the United States, Kohn predicted that "ultimately, [such a] democracy will be determined by its strength as a moral and spiritual factor dominating the public mind."[33]

At the root of legal practices is the concept of *justice*, which has been defined in many ways, from the principles expounded in Plato's *Republic*, through the "state of nature" concept developed in Hobbes' *Leviathan*, to modern interpretations found in the works of ethicists such as Rawls. (Just behavior is, of course, behavior that takes into account the interests of others.) Standing at something of a midpoint between ancient and modern conceptualizations was Hume. His interpretation perhaps represents a summary view that goes to the heart of the issue in its typically sensitive appreciation of the problem.

Hume contended that in a world in which each individual had access to the means of gratifying every passion, there would be no need for justice; for sharing; for property and other rights. "For what purpose make a partition of goods where everyone has already more than enough?.... Why call this object mine when, upon the seizing of it by another, I need but stretch out my hand to possess myself of what is equally valuable?"[34] At the opposite extreme an individual besieged by criminals, and without access to any aid or relief, would, Hume believed, be appropriately excused from ordinary social demands. "His particular regard to justice being no longer of use to his own safety or that of others, he must consult the dictates of self-preservation alone, without concern for those who no longer merit his care and attention."[35]

On the basis of this reasoning, Hume concluded that justice must be recognized as a program for partitioning societies goods while maintaining stable ownership relationships. "Hence, the ideas of property become necessary in all civil society; hence justice derives its usefulness to the public; and hence alone arises its merit and moral obligation."[36] He concluded that "the ultimate point at which they [statutes, customs, precedents etc.] all professedly terminate, is the interest and happiness of human society."³⁷ That is, the "general welfare" should always be taken into account.

An analysis of Hume's position, to the extent that it mirrors that of many philosophers, reveals that it implies—where it does not clearly state--that the principle of altruism is involved. If no one needs assistance altruistic behavior should not be necessary, and in cases of duress it may appropriately be waived, but in normal situations it *should* be applied. However, once again, he reduced its purpose to profit to the individual, pointing out that since "the common situation of society is a medium amidst all these extremes [although we are partial to ourselves and our friends, we are] capable of learning the advantage resulting from a more equitable conduct."³⁸

Legal ethics. Referring to laws in the United States, Boyce and Jensen stated that "the relation between morality and American law is apparent.... The constitution of the United States provides for protection of various rights, apparently on the basis of either natural law or natural rights."³⁹ They went on to say that "statutes are based on the assumption that some things are right and some things are wrong: the foundation for that obviously is a moral one."⁴⁰ Berman took the same position. "By requiring that all laws must conform to...moral principles, the constitution has encouraged American justices to submit to the test of conscience...all legal rules and all governmental acts, including their own judicial decisions."⁴¹

Hill drew a direct relationship. "Ethics precedes and leads to law.... The ethics that is embodied into a law is usually based on the morality of the majority of a society."⁴² He quoted Earl Warren, who, in a 1962 speech said, "In civilized life, law floats on a sea of ethics. Each is indispensable to civilization.... Without ethics law could not exist."⁴³ And, more specifically, Hill defined ethical behavior as honest action. "One cannot be ethical without being honest.... Honesty, being integral to ethics, is part of the earliest survival principle of mankind."⁴⁴

To be honest in one's dealings with another means to be willing to sacrifice part of one's privacy; an altruistic action. The relationship between altruism and law is not, of course, one sided. Those who argue for an anarchistic political system, or for the elimination of some law make their argument on the same grounds; that laws they wish to see proscribed violate the "rights" of those whose behavior such laws restrict.⁴⁵

In a burst of enthusiasm for the meritorious work performed by lawyers, Gambrell, 1956 president of the American Bar Association, writing in The Lawyer's Treasury, said that "as professional men both in England

and America they [lawyers] have lived by their own high code of ethics and their own moral and educational standards. They may well be proud of their record as architects and builders of our free society."[46] Jaworski commented further on the ethical practice of judges and lawyers.

> To serve the public interest, the bar must...achieve a balance of rules that affect a lawyers relationship with other lawyers...with those whom he represents...and with the government that franchises him.... For example the code must resolve...the inevitable conflict arising when the lawyer...decides whether he preserve a client's confidences, or reveal that the client has committed a crime or perpetrated a fraud.[47]

Not everyone among the lay public would necessarily agree with such high sounding praise. Jaworski, for example, felt that the faithful execution of such responsibilities has been seriously compromised by the use of the adversarial system for discovering truth. He quoted Judge Simon Rifkin as saying "the object of a trial is not the ascertainment of [absolute or objective] truth but the resolution of a controversy by the principled application of the rules of the game."[48] In spite of the fact that ultimate truth is not the target, Rifkin proposed that the procedure should (sic) be carried out on the basis of moral principles.

Unfortunately such principles are often subverted. Jaworski pointed out that "to be realistic we must expect some failures and scandals involving lawyers from time to time...regardless of the action of the legal profession to discipline itself."[49] He made a comment, however, that bears on the contention here that laws and lawyers are considered part of a moral system. "If a lawyer does not invoke and practice the moral requisites of the profession," Jaworski said," he dishonors it."[50]

Medical ethics. While the practices of many members of the legal profession are held to be something less than admirable, physicians have traditionally been considered a relatively moral group. Parrot said polls show that "most people consider their physician to be highly ethical."[51] Although such a statement is undoubtedly true for the majority of physicians, the fact that the practice of medicine has become highly lucrative, especially for specialists, has raised doubts about how far such praise should be extended. Furthermore, in America, health care costs have risen dramatically in the latter half of the twentieth century and physicians have received their share of the blame. Insurance costs and the impact of an increasing litigious populace have caused physicians to assume an increasingly defensive posture, often prescribing procedures that may be unnecessary. Parrot contended, however, that when a doctor is "super defensive," the ordering of many batteries of tests "amounts to an economic rip-off of the patient, wholly unethical.... [At one extreme] it becomes an exercise in cowardice...while at the other it is a study in greed."[52]

Although Parrot expressed concern over the possibility that physicians may take too many precautions in ordering medical tests, he was aware of the difficulty involved in determining what should be done in any particular case. "If a given procedure is indeed unnecessary, then there is no question that it is dishonest or unethical. But the whole concept of unnecessary hinges on a definition of what is or is not necessary. A universally acceptable definition...has yet to be devised."[53] He suggested, finally, that physicians should adhere to the scientific method. "It must be used honestly and applied rationally.... There is no middle ground to honest facts, honestly gathered, honestly considered, which result in honest conclusions, and honest human action."[54]

There can surely be no question that moral principles are to be applied, and that in calling for honesty, altruistic behavior is being referenced. In fact, concern regarding the practices of some physicians (as well as that of insurance companies) has resulted in a clamor for the socialization of medical services in the United States, even to the extent of instituting a system of socialized medicine.

Thought Experiments

The self-assertive desire

Self-assertive desires are the most compelling and most apt to be dealt with preferentially. This has resulted in the development of the sociobiological hypothesis of genetically induced selfishness; the conviction that genuinely altruistic behavior is impossible in principle in spite of the palpable evidence. To challenge the validity of that claim, a series of possible behavior sequences or patterns shall be analyzed. In each instance, the costs and wants described are considered to be exhaustive. The cigarettes referred to in the first exercise are, for example, assumed to be easily available, free of charge, pose no health hazard, etc. Recall that all need and cost values are psychological in nature; being determined by the attitude of the behaver. Furthermore, in each example all individuals believe that they have the capacity and opportunity to act.

> Suppose a law were to be passed that prohibited smoking in public buildings. How would the behavior of a man who smokes in spite of

Figure 6.1
Smoking vs Punishment

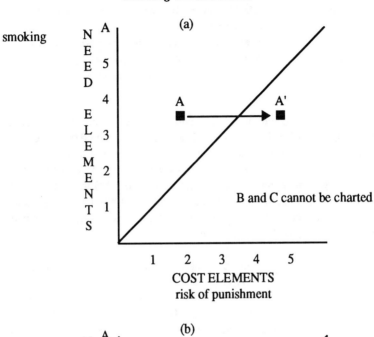

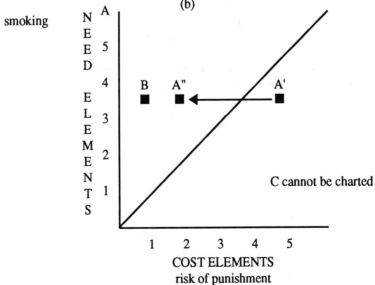

that law be analyzed? The obvious interpretation would be that his desire to smoke is strong enough that he is willing to risk castigation. The need value for that man (A, Figure 6.1a), can be assumed to be greater than the influence of the costs. Presume that at some point, he stops smoking in such places, (A') stating that he no longer wishes to risk punishment. The cost can now be assumed to be greater than the value of the smoking experience. He is willing to give up smoking to avoid any penalty.

Nothing can be said about the desire to smoke of individuals that do not smoke in public buildings. No desire can be presumed on the basis of their failure to smoke. It would make no sense to claim that they do not smoke o*nly* because of the law. The law relates only to opportunity, not to desire. Two other individuals (B and C, Figure 6.1a) may or may not desire to smoke. In this instance the smoking prohibition can be assumed to represent a cost only to the man who smoked earlier and, as has been pointed out, *costs do not alter need values.*

Assume now that the law that forbids smoking is repealed. What might be said about the motive of the man who had quit smoking in public buildings, if he begins to do so again (A" Figure 6.1b)? It would be absurd to contend that the reason that he does so is simply because it is legal. If he smokes, he must *desire* something. In most instances that desire is to enjoy the experience. One other man (B, Figure 6.1b) now smokes a cigarette. A third man (C, Figure 6.1b), does not.

Regarding the behavior of each individual, the most reasonable conclusion would be that the two men who smoke wish to do so, and that the third man probably does not. (Of course, for example in the case of teenagers, the desire of a smoker may be only to impress others, or to experiment with a potentially exciting—if illegal—stimulant, in which case the behavior would be instrumental.) The principle involved is that *every behavior involves some desire.*

If a man says that he bought a shirt because it was cheap, it is most reasonable to assume that he wanted a shirt, and the fact that it was cheap was an incentive to buy *that* one because it was on sale, or perhaps simply because he believed he may have some future use for it. He *wanted* something and was *willing* to pay for it. The reduced cost did not influence his desire. It did provide an opportunity, and perhaps increased his capacity to make the purchase.

The self-protective desire

The same procedure employed with the self-assertive desire can be used to demonstrate the self-protective desire.

> Suppose a young man to be strolling in a pasture, the purpose of the walk (the need value) being to enjoy the day while getting some exercise. The desire involved would obviously be for sensual stimulation. His BAP location would be at A (Figure 6.2a). The cost in terms of time and energy B (Figure 6.2a) as well as the sacrifice of the expression of alternative needs would obviously be minimal.
> Assume that he is suddenly threatened by an angry bull that has wandered into the pasture. He would certainly make every effort to get out of the field. His BAP location would be at A' (Figure 6.2b). Costs (B', Figure 6.2b) include the energy expended and the possibility that he may be driven by the bull into a cul-de-sac, from which he cannot escape.

The gain in such a situation would clearly be greater than the cost. Aware of no alternative possibilities, the young man would be unable to consider any other move. In this case, the "nonbehavior" would be only the continuation of his pleasant afternoon walk, which is identified on the BAP at location B (Figure 6.2a). It should be obvious that the desire for safety is contingent on the presence of a threatening situation when the desire is experienced. The recollection of a dangerous event, although it may bring to mind the fact that one was frightened at that time, does not call for a response designed to cause behavior that will make the individual feel safe. The emotion relates only to the present situation.

The self-transcendent desire

On the basis of the reasoning involved in the determination of the role of desire and its ubiquity as a behavioral element in the case of self-assertive and self-protective urges, behavior directed toward the welfare of others must be interpreted as similarly initiated. The principle of the negative signal, and the behavior associated with its resolution is appropriately applied. The encouragement to assist others must be based on another class of desire, the *transcendent* urge, which represents a concern for the well-being of others and the practice of altruistic behavior.[55] It must be understood as an affective state which encourages gene pool enhancing behavior, although the focus of the resulting behavior can be shown to profit the community directly.

Figure 6.2
Escape vs Injury

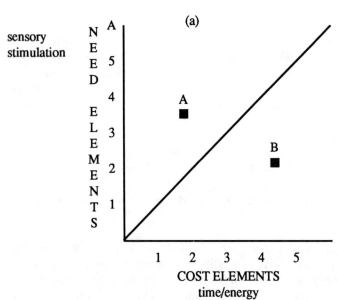

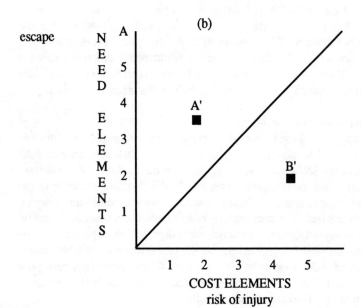

Following are examples of ostensibly altruistic behavior and the transcendent desire on which they are based. Generous, self-denying, behavior is common among people of all levels of affluence. In the case of those of significant wealth, however, it has been argued that the motive is perhaps simply that, following Hume's suggestion, they have a superabundance of whatever they may need. Consider a possible behavioral event among a group of individuals who have unlimited opportunity to indulge every conceivable desire that they may wish to gratify. (A, Figure 6.3a) Ordinarily their behavior would be based principally on assertive desires with any contribution to others (B, Figure 6.3a) representing a sacrifice, however little.

> Suppose a request were made of half of them that some relatively small sum, perhaps $1,000, be anonymously provided to help a hungry child in some distant place. The others would be asked to contribute a similar amount of money to a charitable organization. Following the exercise, donors would be asked to describe their affective state when making the bequest. Most individuals asked for such assistance would probably accept the obligation, and report feeling positive about it (A', Figure 6.3b). Costs (B', Figure 6.3b) would be considered trivial.

It would not make sense to argue that they gave because they would be sacrificing so little—because they had the capacity, and were now offered an opportunity. Just as in the case of the motive for smoking when the law changed, the behaviors indulged cannot be considered to be based on the elimination of a hindrance. Individuals who give under such conditions must desire *something*.[56] For one group, the consequence of the behavior would be known. However, for the other, where no specific information regarding the use of the donated funds was provided, the donation would still qualify as altruistic. In the situation described here, almost all possible costs (e.g., interference with the opportunity to express other needs or other desires) have been eliminated. The elimination of costs, however, is not capable of creating a desire.

If people are willing to assist others when they can freely choose not to, it is appropriate to assume either that they feel the pressure of the transcendent urge, (that they have an interest in the welfare of others), or because they hope for some reciprocity. But in the example cited above, what possible return may such people expect to get from the child that they assist, or the agency to which they contribute? And how can a kin selection factor be assumed to be operating?

Figure 6.3
Altruism vs Egoism

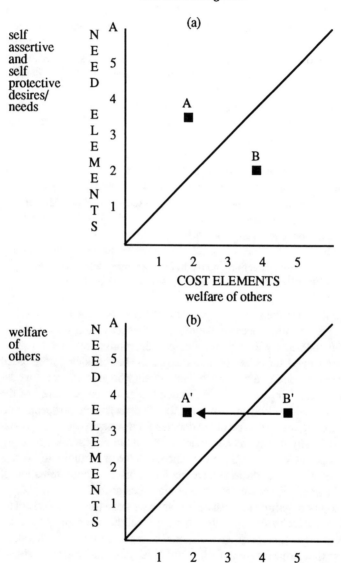

There is, of course, a temptation to argue that those who give do so only because they believe that they *should*; that obligation rather than desire is the motive (I don't want to but I believe that I should.) This is, of course, to ignore the influence of the transcendent urge. Furthermore, when the potential consequences of such behavior are considered, the objection is readily met with the question: Do not such people desire the well-being of those that they assist—whether the specific donees are known or not? Must they not want those that they succor to profit from the process? A negative response would not only be iniquitous, it would manifest a degree of foolishness on the part of the donor.[57]

It would be equally inane to contend that the reason for giving in the egoistic-altruistic situation described above, is that the benefactors have all that they need. Hume's argument for giving was that infinitely wealthy individuals can always replace that which is given. Recall his comment. "Why call this object mine," he said, "when, upon the seizing of it by another, I need but stretch out my hand to possess myself of what is equally valuable?"[58]

But Hume's argument is not applicable where one is *choosing* whether to give. No case can be made for charity that is contingent simply on the occurrence of the opportunity to do so, any more than, as was shown above, it is possible to draw such a conclusion in situations involving self-assertive and protective desires. One must deal with the question of the motive for *giving*, not the lack of a motive for refusing.

In each example above, what happened was that an *opportunity* was provided. The legalization of smoking, and the request for assistance to a hungry child each provided an opportunity. In no case was a desire created (though it may have been brought to the individuals attention.) The issue, as it involves altruistic behavior is complicated by the fact that both protective and assertive desires are more compelling. However, it should be clear from these examples that they cannot be proven to represent the *only* desires. Consider once again several of the characteristics of behavior described in Chapter 4 as they are applied to each instance.

All behavior is based on belief. In the examples above, the most reasonable assumption would be that the individuals that behaved must have believed that smoking, escaping, and altruistic behavior would be positive experiences. In the latter instance, the individuals involved probably also believed that they *should* act as they did. (The possibility that each behavior was instrumental must, of course, be considered.)

Behavior is fully determined by the want/cost ratio. Whether an individual reaches for a cigarette, turns down an offer for a second dessert, or contributes to a charitable organization, the desire can be ascertained by considering each of the three types of motive, and eliminating those that are inappropriate. In the case of the transcendent urge, the desire is accompanied by a sense of the propriety of the behavior. There is nothing "extragenetic" operating. Most importantly, determining the want/cost ratio in every situation is unique. The reasons are idiosyncratic.

Wants and costs are totally independent of each other. An appreciation of this characteristic of behavior is vital to any effort to reveal the existence of a moral sense, and a willingness to act altruistically. In each example above it is obvious that cost alterations did not effect need levels, or vice versa.[59] Concern, of course, is with the impact of altering costs when altruistic behavior is possible. Lowering the cost could not create a desire to sacrifice, any more than the repeal of a law could cause a desire to smoke. What happens in every instance is that as costs are reduced, only desires—heretofore unexpressed because of belief regarding the extent of the costs—can be revealed. Unless such a desire exists, the individual not only would not sacrifice for others, *they could not*!

Altruistic behavior and moral credit

In view of the fact that all behavior is based on a desire, when—if ever—should moral approbation be applied? That question requires an understanding of general behavioral principles. It was pointed out in Chapter 4 that, along with desires, all behaviors involve costs, which represent actual or potential interference with the expression of competing desires or of the relinquishing of alternative needs relevant to the same desires. Thus, the cost of eating may include the probability of gaining weight vs. the desire to be slim, etc. Additionally, the item eaten may be taken in the place of some other food. At least some such situation must be identified when cost is assayed. In the case of acting in the interest of others, where risk or potential sacrifice is recognized, altruistic behavior must be seen as a *cost*; an instrumental behavior.

If an individual has a desire to eat, although a meal is available, many costs must be considered. These include, in addition to any dollar cost, such things as energy and time, as well as the sacrifice of all alternative needs and/or desires. Furthermore, there are usually instrumental behaviors, such as getting to the food, which also represent costs. The same can be said for behavior based on the protective urge where, for example, repairing to a safe place is instrumental.

The assignment of moral credit to altruistic behavior is simply a recognition of the role played by the moral sense. In the performance of self-assertive and self-protective based behaviors, the individual (the inclusive self) receives the ultimate benefit. In the case of behavior based on the transcendent desire, the benefit goes to others. Sacrifices are made so that others, including those not related by blood, may benefit. They are made on the basis of the inherent self-transcendental desire, and a willingness to risk. They are not egoistically motivated. This is perhaps the most subtle of the aspects under consideration.

Consider the tendency of people to feel sorry for those injured or killed in a tragedy; even those who may have been strangers, or perhaps enemies. This widely acknowledged sentiment suggests that one of the costs associated with pernicious behavior is an awareness of a concern for others. It is probably the most significant cause of the restraint that is observed in the carrying out of potentially destructive behaviors. Even the individual who shows little or no compassion for the health and welfare of others is apt to demonstrate a sense of responsibility if, for example, the security of a close friend is threatened.

An altruistic gene?

It is commonly assumed that a gene is a blueprint. But a blueprint is only a physical manifestation of an idea. Presume that a blueprint of a house is created; a document on which is printed a pattern from which the house will be constructed. What happens if the blueprint is lost, although the creator of the blueprint knows what was on it and can proceed without it? The idea of the house lives on. *Is the idea real?* An individual holds a telephone conversation. During that time the delivery of some item from a local store is ordered. The item is delivered. Was the conversation *real? Did it exist?* Or is only the palpable; the tangible; the corporeal entitled to quiddity status? That question has been probed by philosophers for centuries. However, the issue of interest here concerns only the question: *Do ideas have an influence on behavior?*

When sociobiologists speak of the influence of the gene on human behavior, the question to be dealt with is: Where does the information (knowledge) with which the gene is invested come from? The father and mother do not *create* the zygote. Neither they nor their gametes *invent* a program. The fact is that the ideas transmitted come from a random assortment of genes (blueprints) that pervade the gene pool. That which survives over time is the idea on the basis of which individuals come into existence.

Behavior that appears to be altruistic may be interpreted as representing selfishness—an ineluctable self-enhancing response to the demand of the genetic idea. But *must* it be interpreted in that way? Must *all* behavior, animal, or human, be directed toward the ultimate interest of the behaver or the behaver's kin? In point of fact, it may be that what appear to be selfish behaviors—those based on assertive and protective desires—are really *themselves* altruistic!

Recall Dawkins' interpretation of a bird's behavior in a threatening situation. His suggestion is not criticized on the grounds that it presumes a degree of unsuspected cleverness in the bird, since such a capacity may well be genetically induced. There is, however, another way to explain such behavior. One might begin with the description of a bird's seemingly altruistic behavior, and propose that it could just as well represent an example of action geared to the enhancement of a gene pool. Consider the following scenario:

> On observing a bird giving what appears to be a warning cry, a sociobiologist remarks: "What a clever technique that bird has developed for protecting himself. He is using his unsuspecting colleagues as foils. Some of them may die, but the odds favoring his own escape have been dramatically improved."
>
> A holarchic biologist replies: "Not at all. Although it may be true that the bird emitting the cry has improved its own survival potential, creatures at that level of intellect have no notion of what the outcome of their behavior may be. Such intellectual acuity is surely beyond them. The significant factor is that a cry which risks the safety of the individual uttering it is probably part of an adaptive process whereby the flock's survival potential is enhanced. Such behavior is an element in an evolutionarily stable system. The bird, in fact, need be no more aware of its service to the flock than of its own probability of survival." The outcome of such behavior qualifies it as altruistic.

The sociobiologist's response would be that partiality for the former explanation is consistent with what biologists know about the nature and function of the gene and the fitness phenomenon. The palpable evidence of self sacrifice, however, may well provide an explanation that demonstrates an advantage accruing to the gene pool that requires the sacrifice of some members of the group. Statements to that effect have, in fact, been made by many members of the biological community.

There are a variety of processes such as reproduction, for example, which do not serve the interest of the individual directly, if at all. Sociobiologists contend that reproductive activity is pursued because it results in the survival of that individual's genome. In such situations, however, the gene pool or system is also served, the individual being simply a manifestation of that entity. The significance of the practice of generating offspring cannot be overstated. Entropy resistance and differential fitness among individuals which enhances living systems are expressed by way of the process. While a specific lineage may appear to take precedence under one interpretation, the gene pool is, in fact, the system which survives. It does so on the basis of the reproductive success, or fitness of its phenotypes. Though its properties are altered over time, such phenotypes are no more than modifications of those of earlier generations.[60] They are simply manifestations of a gene pool.

This suggests the possibility that even blatantly egoistic behavior may be only "reciprocally" selfish (selfishness practiced in the interest of serving others). One's responding to self-assertive and self-protective desires may be actually a manifestation of a species or gene line adaptive process. In every instance, behavior focusing on such desires may be a function that serves the interest of others, in spite of the fact that the individual may enjoy the experience. Since the fitness of a gene pool is enhanced to the extent that the welfare of its members is accomplished, there is no question that the practice of caring for one's self first has the ultimate effect of providing profit to the gene pool.

When an individual responds to self-serving motivational influences, the source of such activity can be traced to the gene pool from which it emanates. But what is the rationale for such behavior? Why is it propagated through the reproductive process? Why does the individual seek to further its own interests, often to the detriment of others? The notion of an altruistic interpretation of such behavior provides an explanation. If a parent is to care for its offspring efficiently, it must certainly pay attention to its own well being. The wildebeest or zebra mother must develop considerable strength if she is to be successful in fending off predators. She must eat and drink from whatever nutrients are available to develop that ability. Human parents require a similar aggressive capacity if they are to protect, support, and otherwise care for their children.

A football coach applauds his players for eating voraciously, but from his perspective only in order that they may better serve the team. A general makes sure that his troops are well provisioned so that they remain at the peak of readiness. A dance instructor implores her students to watch their weight in order to appear at their best for the pleasure of an audience. There is, thus, often profit to others in action that appears to be motivated only by egoistic urges.

Wilson, in his study of ant and bee colonies, demonstrated that "queens" are fed and protected so that they may perform duties required by the brood. In some colonies, he said, "the queen has regressed in evolution toward the role of a simple egg laying machine."[61] In principle, the assertive desire for food serves the function of providing individuals that are valuable to the gene pool. Similarly, "most of the shelter for the queen...is provided by the bodies of the workers themselves."[62] Thus the protective urge is extended to the individual that has the capacity for producing more members of that gene pool. In each case, behavior that appears selfish can be understood to be related to an altruistic, or other-directed goal!

In social and political systems, individuals are often sacrificed to the welfare of a community. Soldiers, often pampered by a government, risk their lives to defend a country. Programs are developed in order to educate children, because of "a society's need for the best and the brightest." Of course, in many such instances, the individual's behavior meets assertive desires. The soldier eats because it "feels good" to do so. The welfare of the community seems incidental. However, the mother bird that feeds her children is also doing so because it "feels good." In both cases, desire lies at the root of behavior, and the welfare of the gene line is served by the behavior of its phenotypes.

How does the logic of the two positions compare? Dawkins, and other sociobiologists contend that the "proof" of selfish behavior resides in the fact that selfish genes are maximally fit. The "proof" of the hypothesis that the function of self-oriented behavior is ultimately profit to the gene pool, and thus subtly altruistic, is that (a), although under most conditions gene pools do well when their phenotypes prosper, there is incontrovertible evidence that under certain conditions it is most profitable to sacrifice some individuals for the enhancement of the group, and (b), gene pools outlive their members in every case.

Social, Political, and Educational Goals

There are a variety of worthwhile goals involved in the development of a community of individuals that accept the responsibility inherent in effective personal and social relationships. Guisinger and Blatt stated that "The recognition of the importance of both self-development and interpersonal relatedness...can provide a theoretical basis for appreciating and encouraging the development of these essential dimensions in all members of society."[63] It is, of course, equally important that such a "theoretical basis" be translated into practical programs, which requires the appreciation of the fact that effective interpersonal relationships call for the acceptance of a moral sense that influences altruistic behavior.

Beyond the fact that the wide practice of altruistic behavior would be advantageous to a community, is the equally important consideration that through risk and sacrifice *people learn to love*. The pertinent principle is that individuals love most deeply those to whom they give. In practicing altruism people learn to identify with those that they assist. An expanding self emerges. The individual who receives assistance may develop a feeling of appreciation and respect, but love is an affect that develops out of a sense of commitment; of involvement; of obligation.

And how may the behavior of those who do not—or cannot—love be interpreted? Wilson asked: "Are there internal (self-interested) reasons, which make it profitable for [the person] to be thus committed [to an ostensibly loving relationship], reasons that make it profitable for him [to act altruistically] yet reasons which are intrinsic to being committed and not just instrumental?"[64] In a tribute to morality, he proposed that, "the naughty and the selfish are just one case of moral inadequacy. The other... case is presented by the indifferent and the despairing. Both are examples of inability to love, which is why the notion of love turns out to be central."[65]

Warnock contended that "one may say...that most human beings have some natural tendency to be more concerned about the satisfaction of their own wants, etc. than those of others."[66] Wilson, however, argued that loving provides a kind of safety, that it is "a matter of *having a world to live in*,"[67] and that "safety consists precisely in the rejection of autism, in being able to come to terms with and accept the world as it is.... It is only in the company of other people and with their help that we can make any serious assault upon the problems of our own lives. Autism is the chief enemy of becoming good, as it is of being good."[68]

That philosophy; the equating of loving with security, with peace, and the acceptance of others leads to a profit to the individual—an apparently "selfish" outcome. However, the basis for the feeling of satisfaction is in what one does for others; what altruistic behavior is undertaken. The effort to initiate such a life style may seem difficult. Wilson once again offers a suggestion.

> We may be convinced that it is impossible for us, even with our best endeavors, really to love another person, or enjoy a job of work, or even feel at peace in our own home.... [But] once we have taken the first steps in the quest for love, the quest itself becomes interesting,

exciting and lovable. We learn to enjoy becoming good, as well as being good.[69]

Such terms as responsibility, obligation, and commitment, as well as the altruism they engender are essential to such an enterprise. They are meaningful in that they bind individuals to more comprehensive entities—to their fellow human beings. An individual who experiences a sense of responsibility—and thus of love—expresses that feeling in behavior directed toward particular groups; toward organizations or communities which the individual invests with value. Once again the distinction between *ethics* and *morality* must be appreciated. The capacity to feel obligation is inherent; the expression of that feeling is learned.

Values can, of course, be taught if one means by that the teaching of who or what are appropriate meritorious objects, as when a mother teaches her children to respect a teacher, or a community cleanliness code. In many instances, however, individuals find reason to respect institutions, or other individuals without formal training. In all such instances the notion of *community*, described earlier, is involved; where a distinction is drawn, and commonalities recognized, between *self* and *other*.[70] Certain basic problems persist. If one accepts the holarchic principle that individuals desire to assist others, why should people be rewarded for doing what they "want" to do?

First because it is desirable that individuals recognize the instrumental nature of giving, and the rewards involved. It is essential that children learn to read, so they are rewarded for studying; to be slim so they are rewarded for dieting. In such cases, it is assumed that the individual will appreciate the outcome. It is also critical that children learn the *intrinsic* satisfaction related to giving or sharing.

It is because of the perplexity of such questions that the concepts of ethics and morality become confused. It is why many theorists, in their failure to find universal ethical principles, deny the legitimacy of a moral sense. The confusion can be addressed however, by accepting the general principle that altruistic behavior is *universally* believed by civilized people to be a moral practice. It is born of an innate sense of responsibility that is satisfied by the practice of "good works," and the loving feeling that such behavior creates.

The opportunity to share should be encouraged at an early age. Unfortunately, the approach to developing altruistic behavior in children is complicated by the nature of the social system. Wilson made the point that it is essential to teach children to appreciate the value of practicing altruistic behavior. As Wilson said, they cannot be expected to give up [their] original objects of desire because they are forbidden or incapable of attainment. They must learn that altruism is inherently rewarding. They must be educated in the act of social living.

Such education should, of course, be initiated in the home, where opportunities, often neglected, abound. "Rights" may be replaced—or at least supplemented—by "obligations"[72]; opportunity to give may sometimes be instituted where one usually receives; sharing may replace sole ownership; "we" may be employed instead of "me." There is certainly an essential role for self-oriented action. The self-assertive and self-protective urges are extremely important. However, the practice that has developed in America over the past several generations has been to model the behavior of the "takers." That augurs a tragic future for democracy, and *love is one of the emotions that has suffered in the process.*

A Recapitulation

Three interpretations of other directed behavior have been presented.

- Culturists claim that genuinely altruistic behavior exists, but that it is learned, and is more a product of social than genetic factors.
- Sociobiologists contend that genetic fitness demands exclusively selfish behavior. They claim that genuine altruists cannot survive reproductive competition.
- The "holarchic," or gene-pool-evolutionarily-stable-system demonstrates that the genome is the agent of gene pool initiated urges that include altruistic as well as selfish motives.

None of these models denies that altruistic behavior occurs within the context of family membership, although it is, of course, quite conceivable that physical proximity and interactive experiences cause bonding to occur that accounts for the tendency to prefer particular groups and individuals. Nevertheless, the notion of "inclusive fitness," which refers to the existence of an influence that encourages assisting one's relatives in descending order as the degree of kinship fades may be an accurate interpretation. Unfortunately, the rationale for such behavior, as proposed by biologists, is limited to suggestions regarding possible causes.

Culturists come in two varieties. One group accepts the "selfish gene" hypothesis but contends that it is supplemented by social protocols. Although genetic influence is important, people can, they believe, "adopt a second set of ethical norms" as was discussed above. The second group is comprised of hard core environmentalists, who believe that all behavior is a product of social forces. They insist that, "cultural explanations [for altruistic behavior] are readily available."[73] It is argued, for example, that "making friends" and indulging in similar socially approved behavior helps to "temper predatory instincts."

Irresponsibility and selfish behavior are explained as the products of cultural failings. There is general agreement that altruistic behavior is based primarily on the needs of a society for order and responsibility. And, as is stated here, they contend that an observation of human behavior reveals many examples of generosity and self-sacrifice. Unfortunately, culturist arguments seem to represent no more than a desperate effort to deny sociobiologist's claims.

Although there is a conviction that genuine altruism exists its supporters have been unable to provide evidence to substantiate their claim. They are reduced to insisting that humans have souls; that they possess a sense of propriety; a willingness to aid others in times of trouble. They refuse to accept sociobiological interpretations on the grounds that such explications are "myths," and, according to some even "hellish creations." Butler, Hume, Waddington, and many others claim that there is a principle of "benevolence" that marks off humans as unique. McShea, and others say that the evidence provided by sociobiologists does not support their contention that behavior is wholly genetically controlled.

The sociobiological argument is based on several factors. First, it is assumed that recently acquired genetic knowledge leads biologists to the inexorable conclusion that selfishness is an essential characteristic of all behavior. An evolutionarily stable system is based on the reproductive advantage of the lineage whose members act most consistently in their own interest. Second, observed behavior includes extensive evidence of selfishness. Greed and avarice are found everywhere. Third, what appear to be altruistic behaviors can be equally well interpreted as selfish. ("Scratch an altruist and watch a hypocrite bleed.")

The holarchic model takes issue with each of these positions. The sociobiologist argument represents no more than a rational interpretation, with no proof of any aspect being provided. The principle of "reciprocal altruism" represents no more than a proposed interpretation of sacrificial behavior that reconciles it with the demands of the "selfish gene" hypothesis. Although the theory that cooperative behavior will evolve as an outcome of the behavior of individuals "pursuing their own self interest," is correct, it is not the only manifestation of an evolutionarily stable system. Such a system may ostensibly function to enhance the fitness of a group such as a gene pool, as individuals risk and sacrifice. The cooperative strategy is in fact, however, apt to be ESS negative, since it is a manipulative technique for taking advantage of others. The individual's interests are paramount.

The holarchic model includes the acceptance of the notion that self-serving behaviors are important, common, and ordinarily of primary concern. However, the contention that genuinely altruistic behaviors do not exist, and that *all* self sacrifice is based on the anticipation of some future profit to the individual is categorically denied. The conviction that self-denying behaviors unrelated to reciprocity are willingly practiced, is based on several principles:

- Since each of the senses has evolved because it performs a *function*, the willingness to sacrifice for others provides evidence of both the transcendental urge and a moral sense; the conviction that one should accept the obligation to sacrifice in the interest of others—a sense of responsibility to perform behaviors that serve gene pool interests.[74]
- Since every motivational sequence includes some type of desire, altruistic behavior, which represents the widely observed practice of risk-taking and self-sacrificial activity must represent the embodiment of a desire for behavior that is directed toward the welfare of others, whether or not the consequence of such behavior is known.
- A research hypothesis that states that "some x's are y's" is considered to be proven when one x meets that condition. One example of genuine altruism (which every significant sociobiologist concedes to exist) provides the proof that is required.

Evidence of many such behaviors is readily available:

- Most people are willing to act generously when the more demanding self-assertive and self-protective desires are met.
- Extensive examples of altruistic (as well as egoistic) behavior are observed in every part of the world.
- Behaviors proclaimed to be "reciprocally" altruistic by sociobiologists may just as convincingly be demonstrated to be genuinely altruistic.

Summary

Sociobiologists have painted a portrait that depicts all behavior—and particularly that of humans— as ultimately self-serving. Unfortunately a stain has appeared on the surface of that canvas that has persisted in spite of every effort at eradication. An army of professionals—as well as millions of lay people—insist that genuine altruism exists; that along with all of the other human desires, is one for the welfare of humankind. Every effort has been made to "correct" that flaw on the basis of scientifically derived principles.

There was first the contention that only relatives are treated deferentially (Hamilton, 1964, a, b) on the basis of the genes shared by donor and donee. Although this certainly seems reasonable, the immediate *or efficient cause* for such preferential behavior, especially where it calls for self-sacrifice has been shown here to be a function of the motivational process. Furthermore, considerable beneficence is observed that involves non-relatives; behavior that suggests the existence of a feeling that leads to behavior that should diminish fitness.

That problem is addressed by the reciprocal altruism thesis (Trivers, 1971) which is based on the contention that since the gene is the unit of evolution, aiding others must be interpreted as representing a circuitous route to self-enhancement. This explication has been hailed by biologists and ethologists as a brilliant revelation. It fits the requirement of genetic selfishness. But every interpretation of ostensibly altruistic behavior can be equally well interpreted as genuine. The behavior of the jay, as was exemplified above, may be construed as an effort to help other birds at personal risk. The weakness of such an inversion is, of course, that the altruistic individual reduces the probability of reproducing itself. Such behavior should not persist across generations. But gene pools provide the genetic inheritance, and, thus, profit from the admixture of egoistic and altruistic behavior.

Along with the principle of reciprocal altruism came the "manipulation" explanation for what appears to be generous behavior. Recall that Williams argued that "anyone who makes an anonymous donation...is biologically as much a victim of manipulation as the snapper in the jaws of the anglefish."[75] Donors are apparently assumed to be fools, because they allow themselves to be "used." Recipients are "hustlers", who take advantage of people's gullibility. So much, he said, for what appears to be altruistic behavior. Still the conviction that genuine altruism exists went on.

A weapon was finally brought to bear that was believed capable of settling the issue. Smith's introduction of the evolutionarily stable strategy hypothesis (ESS) seemed qualified to eliminate the last vestiges of resistance. While the model does allow for a certain amount of putative altruism, the purpose of such behavior is, once again, said to profit the behaver. Dawkins contended that under the ESS model "the best strategy for an individual depends on what the majority of the population [each one trying to maximize his own success] are doing."[76] He pointed out that although some behavior appears unselfish it isn't really so. "Superficially," he said, "this sounds a little like group selection, but it is really nothing of the kind."[77] He proposed that it is no more than a technique that characterizes an evolutionarily stable system with the individual's fitness as the goal.

The ESS principle was employed by Axelrod to demonstrate that individuals "pursu[ing] their own self interest," could learn to cooperate, "without the aid of a central authority."[78] This would seem to close most, if not all, doors. No matter what sort of seemingly generous behavior is observed, he argued that the goal can be shown to always be individual enhancement. In this text, it has been demonstrated that the evidence does not support a purely selfish hypothesis; that humans, as well as many animals, manifest a desire to provide assistance to others, although in so doing they often risk their own welfare—sometimes even their life.

An analysis of altruistic behavior illustrates that the latter interpretation is more reasonable than that of sociobiologists. When attention is paid to the evolution of the motivational process it becomes evident that all behavior can be characterized as a function of the relationship between desires, needs, and associated costs. The type of evidence for altruistic behavior which is based on the holarchic motivational process is given credence by the behavior of millions of people worldwide, who choose to give or share, which supports the contention that genuine altruistic behavior exists, and that it is the enhancement of a prototype or archetype that drives much behavior.

Chapter 6 Notes

1. Dobzhansky (1970), p. 351. Erlich & Holm, however, claimed that the notion of species, a concept so long tacitly accepted, was coming under attack. "In recent years a group of biologists has questioned the utility of the biological concept.... They see it as an artifact which cannot be supported by the evidence—which is that interbreeding seems to be occurring.... The very nature of the biological species definition makes its use in practice impossible," (1963, pp. 299-300).
2. Rosenberg (1985), p. 188
3. Mayr (1988), pp. 314-315
4. Mayr pointed out that species maintain constancy across time through genetic processes that insulate interbreeding populations. Isolation mechanisms, he said, "are a protective device for well integrated genotypes. Any interbreeding between different species would lead to a breakdown of well-balanced, harmonious genotypes, and would quickly be counteracted by natural selection," (1988, p. 319).

    Smith offered data to support Mayr's view. "Perhaps the most convincing argument for the reality of species is that the kinds of animals and plants recognized by pre-literate peoples have been found to correspond almost exactly to the species recognized by modern taxonomists in the same area. If species were merely arbitrary groupings adopted to make naming possible, this would not be so," (1986, p. 41). His conclusion has been accepted by those calling themselves *essentialists*; those who believe that each species has certain essential qualities which are found in each of its members. As to higher categories, Smith felt that a certain hesitation is appropriate in considering them real, in spite of the fact that in some instances the comparison with pre-literates holds.
5. These systems provide the potential for the development of meaning that defines the relationship between a phenotype and a gene pool. The meaning of a gene pool is a characteristic of its relationship to a species. See Wonderly (1991, pp. 50-55) for a discussion of the concept of meaning.
6. Warnock (1971), p. 166
7. Wilson (1975), p. 121
8. *Ibid.* Wilson pointed out that though vertebrates rarely commit suicide, "many place themselves in harms way to defend relatives," (*Ibid.*).
9. Rachels (1991), p. 150. A series of controls were employed to eliminate a variety of possible alternative explanations. Such possible factors as the gender of the subjects, dominance patterns that may call for subservience, and previous experience as a subject were carefully controlled.

10. *Ibid.* p. 151
11. Kummer (1980), p. 43
12. There are many problems with sociobiological explanations of animal behavior. Birds and animals that give warning cries do not always fly away or otherwise leave the scene before their neighbors. The killdeer in its feigned injury and the baboon in its wild gesticulation are obvious examples of exceptions. There is, moreover, the risk that a potential predator may be too far away, or otherwise so preoccupied, that the warning call will increase the threat to the group. In such a situation, the effort of the caller would be not only useless, it may actually prove counterproductive. Specific evidence comes from research with Vervet monkeys, who give three distinct alarm calls (each of which puts the caller at risk) depending on the form of predator.

    In one instance, researchers reported that a Vervet, on seeing another about to be attacked by an eagle, gave a call ordinarily used to indicate that a leopard was in the area. "The 'wrong' call sent the individual running for the trees.... [It] survived the attack, but would not have done so had an eagle alarm call [to look up at the sky] been given and the appropriate response followed," (Leakey 1992, p. 242). The contention here is that the caller was invested with some genetic signal that encouraged a *willingness* to perform other-directed behavior, and some method of determining which was the appropriate form of warning.
13. Dawkins (1982), p. 239
14. The answer can be sought in two places. First, what has been the result of families concentrating on the well-being of their relatives? What have feuds and clan wars taught us? On a larger scale, what are the outcomes of racial, ethnic, and even patriotic allegiance as they are viewed from the vantage point of nonparticipants? Christian Americans decry the Arab-Jew confrontation, while the Irish internecine struggle is considered a tragic error in the eyes of the billions not involved. How can such paradoxes be explained?

    When one is not personally involved, caring for another—or others—is considered to represent the most creditable expression of morality. Leaders of all the principal religions preach this common theme. There are, of course, innumerable examples of individuals who claim that families should take precedence. And, there is a temptation to accept that notion, since it seems that the alternative is to prefer others over one's immediate family. The morally appropriate practice, however, would seem to be to treat all with the respect due to human life, and to recognize the sanctity of each entity when conflict between levels is unavoidable.

A test of the appropriate focus of moral behavior would be the question: Which kind of behavior would be most apt to satisfy the desire of most people; that where the family is the apex, or that which sees all people—people in general—as worthy of prior respect? Acceptance of the latter view may suggest that a communist, or at least socialist, society would be best for the majority. Unfortunately, the strong sense of self that humans have developed has made every attempt to develop such political entities subject to the power urge of those who come to represent the ruling class.

15. Mayr (1988), p. 27. The "instinctive altruism" to which he refers is based on the moral sense.
16. Warnock (1971), p. 13
17. Erlich & Holm (1963), p. 287
18. Mayr said, "some authors seem to think that all human ethics is more or less inclusive fitness altruism. Other[s] think that when genuine human ethics evolved, it altogether replaced inclusive fitness altruism," (1988, p. 76). He proposed a middle ground, agreeing that human altruism emerged from inclusive fitness altruism, but that "[human] ethical behavior is based on conscious thought that leads to the making of deliberate choices," (*Ibid.* p. 77). That is, while some animal behavior appears to be moral, as are the warning cries exhibited by members of threatened flocks of birds, such actions, he said, are merely instinctive, and are not based on any choice made by the individual. Furthermore, solitary animals "have no behavior that natural selection could convert into altruism," (*Ibid.* p. 76).
19. Guisinger & Blatt (1994), p. 107
20. *Ibid.*
21. *Ibid.*
22. Zahn-Waxler & Radke-Yarrow quoted in Bales (1983), p. 15
23. Aresty (1970), p. 10
24. *Ibid.*
25. *Ibid.* p. 12
26. *Ibid.* p. 13
27. Ford (1988), pp. xiii-xiv
28. *Ibid.*
29. *Ibid.*
30. Durant (1944), p. 393
31. Plato referenced in Thilly (1936), p. 71
32. Copleston (1962), p. 92. Hobbes contended that "civil authority and law [is] the foundation of morality," (Albert & Denise, 1988, p. 129). He took the position that "it is manifest, that the measure of good and evil actions,

is the civil law; and the judge and legislator, who is always representative of the commonwealth," (Hobbes, 1962, p. 310). Baier was less generous, saying that the morality of States ends at their borders. In the case of international relationships, he stated that: "States pay only lip service to morality. They attack their hated enemy when the opportunity arises," (Baier, 1958, p. 313).

33. Kohn (1966), p. 191
34. Hume quoted in Aiken (1970), p. 185
35. *Ibid.* p. 188
36. *Ibid.* p. 189
37. *Ibid.* This reflected the view of many moral philosophers. Shaftsebury, for example, argued that the justice of an action is determined by "whether or not it promotes the general well being," (Shaftsebury referenced in Oldscamp, 1970, p. 155).
38. Hume referenced in Oldscamp (1970), p. 189
39. Boyce & Jensen (1978), p. 235
40. *Ibid.* p. 236
41. Berman (1961), p. 225
42. Hill (1976), p. 4
43. *Ibid.* p. 5
44. *Ibid.*
45. The right to publish and disseminate materials deemed by many to be pornographic is a case in point. The Supreme Court of the United States has been required to deal with the question of whether the right to publish, or the right of the public to protection from the possible effects of obscene material should take precedence. The issue exemplifies the desire to control publications mentioned above. Justice Harlan took the position that although the reading of such "obscene" literature may not be a proximate cause of sexual misconduct, the State may reasonably conclude that over time, the dissemination of such material will have an eroding effect on moral standards.

    His opponents, who did not accept such a long view won the case—but also on a (moral) *right*. The right of an individual to publish. And what of civil issues? "Some argue, for example, that in a crime, but not in a tort, moral culpability is essentially involved," (Laird, 1970, p. 4). However, the issue is not settled by ruling out moral responsibility. The case is (hopefully) settled on the grounds of fairness—though that "fairness" may be spelled out in a civil code that some litigants may question as being what it purports to be.
46. Gambrell quoted in Jaworski (1976), p. 189

47. *Ibid.* p. 181
48. *Ibid.* p. 182
49. Jaworski (1976), p. 192. He gave an example from a well known calamity that involved members of the White House legal staff. "One of the great tragedies of Watergate...was the involvement of an inordinate number of lawyers who engaged in immoral conniving, conspiratorial scheming and just plain unprincipled conduct," (*Ibid.*).
50. *Ibid.* p. 194
51. Parrot (1976), p. 161
52. *Ibid.* p. 166
53. *Ibid.* p. 171
54. *Ibid.* p. 172
55. The evidence for the existence of such a desire is based on the same data that explains other desires. Individuals report such a feeling. For example, on observing an elderly woman being struck by a car, people may state that they want to assist her. (This is not to guarantee that they will provide help, but only that they report feeling such a desire.)
56. This point was recognized by philosophers such as Hume, who, in discussing the possible acquisition of a watermelon, said: "No amount of knowledge [capacity and/or opportunity] is alone sufficient to lead me to do anything as long as I do not want any watermelon. If I do not want to eat it, or give it to someone else, or to use it as a paperweight or for anything else, then all the knowledge I admittedly have will not lead me to try to get it." (Hume quoted in Stroud, 1977, p. 157).
57. Putative exceptions of this type are commonly expressed. People paying a fine for overtime parking say that they "must" pay it, but do not wish to do so. However, in such cases what they desire is to avoid incarceration (self-protective). The payment is instrumental. In the example cited here, one may argue that the only reason for giving the $1,000 donation is "to get rid of the person making the request." In this instance, of course, the individual is not being altruistic, but the behavior still rests on a desire; "to get rid of the person making the request," and the belief that a donation will cause that to happen.

    Take the case of being asked to contribute to the defense of an individual that one detests, even a potentially dangerous person. The only appeal may be that it is a tenet of Christianity that one assist *anyone* in distress. The proper interpretation of such behavior is that the person *desires* to act in accordance with Christian principles. There are no exceptions to the rule. *All behavior is based on desire.*
58. Hume quoted in Aiken (1970), p. 185

59. It is well known that as costs change, efforts are made to reduce the dissonance created by altering ones beliefs. Thus, the "sour grapes" and "sweet lemon" allegories. This is accomplished, however, by substituting alternative needs—not merely because of a change in the costs involved.
60. Loomis suggested that even the blue green bacteria "are probably all descendants of the first effective photosynthetic organism that arose about 3.5 billion years ago," (1988, p. 52).
61. Wilson (1975) p. 433
62. *Ibid.* p. 425
63. Guisinger & Blatt (1994), p. 110
64. Wilson (1987), p. 60
65. *Ibid.* p. 57
66. Warnock (1971), p. 21
67. Wilson (1987) p. 63
68. *Ibid.* p. 64
69. *Ibid.* p. 125
70. But that difference is often only a matter of focus. As a member of a city or state, I want my political representatives to bring me (us) as many benefits as possible. As an American I want my nation to stand above others. When my high school baseball team competes with the one at the other end of town, they are the "enemy." But when that other team wins the county championship I root for them to win at the state level. Where, then, does altruistic behavior become appropriate? If "charity begins at home" where (and when) is home? Or am I equally responsible for the welfare of all people—and perhaps, to a lesser extent, to all of animal kind?
71. Wilson (1987), p. 125
72. An example of the opportunity to replace rights with responsibility or obligation may be observed in the 1994 insistence by Muslim students in France, that they have a "right" to wear head scarves to school that proclaim their religious affiliation. How much less confrontational would be an argument to the effect that they felt an "obligation" to their church to dress in such a fashion. How could that claim be challenged?
73. Solomon (1980), p. 271
74. The determination of which "others" should be assisted is, however, a product of cultural factors.
75. Williams (1989), p. 193
76. Dawkins (1976), p. 74
77. *Ibid.* p. 77
78. Axelrod (1984), p. 6

Glossary

Adaptation Any change in the structure or functioning of an organism that makes it better suited to its environment; i.e., that improves its fitness.

Affect Feeling states that represent knowledge as a mental experience. They include desire, emotion, conviction, and belief.

Agent An entity, in its performance of a function, that serves wholes or systems to which is subordinate.

Behavior Deliberate action based on decisions arrived at through a motivational sequence.

   Innate Behavior which is genetically programmed.

   Incident A simple, complete behavioral action.

   Instrumental Deliberate action performed in order to facilitate some other action or outcome.

   Learned Behavior which is developed through experience.

   Pattern Typical, but not continuous, deliberate activity.

Sequence A series of behavioral incidents, including all related activity.

Cognition The processing of information, including conceptualization, abstraction, and generalization.

Communication The transmission of information in which either the sender or receiver performs a function in order to deliver a message.

Culturalism The conviction that environmental or other nongenetic factors are the principal determinants of moral behavior.

Decision An affective state representing the consequences of weighing the relative merit of competing behavior potentialities.

Desire The psychological manifestation of a drive.

> **Self-assertive** Gratification is based on the assuaging of physical or psychological deficit related to the biological or inclusive self.

> **Self-protective** Gratification is based on order or security.

> **Self-transcendent** Gratification is based on the experiencing of a relationship with a more comprehensive entity.

Emotion The affective experience resulting from the interaction of desires and related needs and the perceived environment, real, imagined, contemporaneous, anticipated, or recalled.

Ethics Codes of conduct established on the basis of the moral sense.

Gene A unit of heredity composed of DNA. Genes possess the knowledge essential to both self replication and the creation of proteins that control developmental factors.

Gene pool All of the genes and their various alleles that are present in a population of plants or animals.

Genome All of the genes contained in a single set of chromosomes.

Holarchy A term which reflects the independence that characterizes the function of lower level echelons of organic systems.

Holon A term which describes all existents at all levels, since they are neither whole nor part in an absolute sense, but are always an aggregate of parts or individuals in one sense and parts or instances of larger wholes in another. Existents may be described as holonic, in that they represent both aspects of existence.

Identification Recognition of the biological self as related to a more inclusive existence.

Interactor Hull's interpretation of Dawkins' "vehicle".

Intuitionism The notion that certain actions or types of actions are known to be right or wrong intuitively (regardless of consequences).

Knowledge The capacity to experience, interpret, and respond to information.

    Instinctive The capacity to experience an anticipatory emotion on the presentation of stimuli not previously encountered.

    Learned The capacity to experience an anticipatory emotion as a result of a learning experience.

Life The form of chemico-physical matter that manifests entropy resistance and continued existence across time through the interaction of cell and cell group, genotype and phenotype, species and genus, biomass and ecosystem.

Morality An innate sense of propriety against which situations, behaviors, and events are evaluated.

Motivation A process that flows continually from any emotional state through deliberate activity which is directed toward the maintaining or optimization of an affective state.

Need A behavior, situation, entity, or event that is believed capable of gratifying a desire.

Netcost The sum of all costs involved in a deliberative sequence.

Netwant The sum of all desires and needs involved in a deliberative sequence.

Provincialism The denial of the validity of any analysis that presumes that living beings are characterized by intuitive powers.

Replicator A molecule having the ability to create copies of itself.

System A collection of interdependent parts and/or instances and the wholes to which they are related.

> **Evolutionarily stable** The principle that a group will enhance its fitness if its members act as though they are aware that in the long run they will profit most from cooperative behavior.

> **Inorganic** Those in which relationships are described in the language of chemistry and physics, and where the dynamism is a force field. Parts and wholes are of equal significance.

> **Organic** Those in which the interests of parts and instances are subordinate to those of wholes. To some extent, part and wholes functions to enhance the whole.

Self An affective status condition which represents both the *cause* and *result* of the interplay of genetically fixed dynamisms and environmental opportunity.

Sociobiology The philosophy that genes are most likely to be represented in ensuing generations if they survive to produce offspring. It is the basis for the "selfish gene" hypothesis.

Vehicle Organisms which transport replicators.

Whole The summative aspect of a holon.

Willing A motivating element referring to the acceptance of costs in the interest of gratifying some desire.

References

Abeles, R., Frey, P. & Jencks, W. 1992. *Biochemistry*. Boston: Jones & Bartlett.
Aiken, H. 1970. *Hume's moral and political philosophy*. Darien, CT: Hafner.
Albert, E. & Denise, T. 1988. *Great traditions in ethics*. Belmont, CA: Wadsworth.
Alland, A. 1971. *The human imperative*. New York: Columbia University Press.
Alper, J. 1978. Ethical and social implications. In *Sociobiology and human nature*. Edited by M. Gregory, A. Silvers & D. Sutch, 195-212. San Francisco: Josey-Bass.
Ardrey, R. 1966. *The territorial imperative*. New York: Dell.
Aresty, E. 1970. *The best behavior*. New York: Simon & Schuster.
Ashman, C. 1973. *The finest judges money can buy*. Los Angeles: Nash.
Axelrod, R. 1984. *The evolution of cooperation*. New York: Basic Books.
Axelrod, R. & Hamilton, W. 1989. Game theory and biology. In *From Gaia to selfish genes*. Edited by C. Barlow, 137-143. London: Massachusetts Institute of Technology Press.
Ayer, A. 1956. *The problem of knowledge*. Baltimore, MD. Penguin Books.
Badcock, C. 1991. *Evolution and individual behavior*. Oxford: Basil Blackwell.
Baier, K. 1958. *The moral point of view*. Ithaca, New York: Cornell University Press.

_____. 1964. Reasoning in practical deliberation. In *The moral judgment*. Edited by P. Taylor, 277-296. Englewood Cliffs, NJ: Prentice-Hall.
Bales, J. 1983. Research traces altruism in toddlers. *APA Monitor*, 15: 20-22.
Balzac, H. 1994. *Pere Goriot*. Translated by B. Raffel. New York: W. W. Norton & Company.
Barash, D. 1977. *Sociobiology and behavior*. New York: Elsevier.
_____. 1979. *The whisperings within*. New York: Harper & Row.
Barnhart, R. ed. 1986. *Dictionary of science*. Maplewood, NJ: Hammond.
Batson, C. 1990. How social an animal? *American Psychologist* 45: 336-346.
Berman, H. 1961. Philosphical aspects of american law. In *Talks on American law*. Edited by H. Berman, 221-235. New York: Vantage Press.
Bernal, J. 1967. *The origin of life*. Cleveland: World.
Bertalanffy, L. 1952. *Problems of life: An evaluation of modern biological thought*. New York: Wiley.
Boden, M. 1984. Artificial intelligence and biologic reductionism. In *Beyond neo-Darwinism. An introduction to the new evolutionary paradigm*. Edited by M. Ho & P. Saunders, 318-329. London: Academic Press.
Bohm, D. 1969a. Further remarks on order. In *Towards a theoretical biology: 2. Sketches*. Edited by C. Waddington, 41-60. Chicago: Aldine.
_____. 1969b. Some remarks on the notion of order. In *Towards a theoretical biology: 2. Sketches*. Edited by C. Waddington, 18-40. Chicago: Aldine.
Boyce, W. & Jensen, L. 1978. *Moral reasoning: A psychological-philosophical integration*. Lincoln, NE: University of Nebraska Press.
Brandon, R. & Burian, R. eds. 1984. *Genes, organisms, populations*. Cambridge: Massachusetts Institute of Technology Press.
Brill, A. 1938. *The basic writings of Sigmund Freud*. NY: The Modern Library.
Broad, C. 1965. The traditional problem of body and mind. In *Reason and responsibility*. Edited by J. Feinberg, 209-217. Belmont, CA: Dickenson.
Burgess, R., Kurland, J. & Pensky, E. 1988. Ultimate and proximate determinants of child maltreatment: Natural selection, ecological

instability, and coercive interpersonal contingencies. In *Sociobiological perspectives on human development*. Edited by K. MacDonald, 293-319. New York: Springer-Verlag.

Butler, J. 1873. *Sermons*. New York: Carter.

Campbell, R. 1980. Social morality norms as evidence of conflict between biological human nature and social system requirements. In *Morality as a biological phenomenon: The presuppositions of sociobiology research*. Revised edition. 47-80. Berkeley: University of California Press.

Caporael, L., Fawes, R., Orbell, J. & van de Kragt, A. 1989. Selfishness examined: Cooperation in the absence of egoistic incentives. *Behavioral and Brain Sciences* 12: 683-739.

Carty, J. 1971. *An introduction to the behavior of invertebrates*. New York: Hafner.

Casti, J. 1992. *Reality Rules II*. New York: Wiley and Sons.

Cattell, R. 1972. *A new morality for science: Beyondism*. New York: Pergamon Press

Charlesworth, W. 1988. Resources and resource acquisition during ontogeny. In *Sociobiological perspectives on human development*. Edited by K. MacDonald, 24-77. New York: Springer-Verlag.

Copleston, F. 1962. *A history of philosophy: Vol. I, Greece and Rome, Part II*. Garden City, New York: Image Books.

Crick, F. 1966. *Of molecules and men*. Seattle: University of Washington Press.

Dawkins, R. 1976. *The selfish gene*. London: Oxford University Press.

———. 1982. *The extended phenotype*. London: Oxford University Press.

———. 1989. *The selfish gene*. New Edition. London: Oxford University Press.

———. 1995. God's utility function. *Scientific American* 273: 80-85.

Day, W. 1979. *Genesis on planet earth*. East Lansing, MI: The House of Talos.

Degler, C. 1991. *In search of human nature*. New York: Oxford University Press.

Dobzhansky, T. 1970. *Genetics of the evolutionary process*. New York: Columbia University Press.

Draper, P. & Harpending, H. 1988. A sociobiological perspective on the development of human reproductive strategies. In *Sociobiological perspectives on human development*. Edited by K. MacDonald, 340-372. New York: Springer-Verlag.

Durant, W. 1944. *The story of civilization: Part III. Ceasar and Christ.* New York: Simon & Schuster.
———. 1950. *The story of civilization: Part IV. The age of faith.* New York: Simon & Schuster.
Dyson, F. 1985. *Origins of life.* London: Cambridge University Press.
Edwards, P. ed. 1967. *The encyclopedia of philosophy.* New York: Collier MacMillan.
Eigen, M. 1992. *Steps towards life: A perspective on evolution.* Oxford: Oxford University Press.
Eigen, M., Gardiner, W., Schuster, P. & Winckler-Oswatitch, R. 1981. The origin of genetic information. *Scientific American* 244: 88-118.
Ehrlich, P. & Holm, R. 1963. *The process of evolution.* New York: McGraw Hill.
Feder, D. 1993. New "rights" are created at a dizzying pace. *The Conservative Chronicle.* 8 December, p. 21.
Flew, A. 1979. *A dictionary of philosophy.* New York: St. Martin's Press.
Ford, C. 1988. *Etiquette:* New York: Crown Publishing.
Fox, R. 1989. *The search for society: Quest for a biological science and morality.* New Brunswick, MA: Rutgers University Press.
Fox, R. F. 1988. *Energy and the evolution of life.* New York: Freeman.
Fox, R. L. 1989. *Pagans and christians.* New York: Knopf.
Fox, S. 1980. Metabolic microspheres. *Naturwissenschaften.* 67, 378-383
Franklin, B. 195? *Benjamin Franklin's wit and wisdom.* Mount Vernon, NY: Peter Pauper Press.
Freedman, D. 1979. *Human sociobiology: A holistic approach.* New York: The Free Press.
Freud, A. 1949. Aggression in relation to emotional development: Normal and pathological. In *Psychoanalytic study of the child.* Translated by A. Freud, H. Hartman & E. Kris. Vols. 3-4: 37-43. New York: International University Press.
Freud, S. 1961. Civilization and its discontents. In *Standard edition of the complete psychological works of Sigmund Freud.* Edited by J. Strachey, Vol. 21, 64-148. London: Hogarth.
Fromm, E. 1973. *The anatomy of human destructiveness.* New York: Holt, Rinehart & Winston.
Ghiselin, M. 1974. *The economy of nature and the evolution of sex.* Berkeley: University of California Press.
Gough, J. 1957. *The social contract.* London: Oxford University Press.

Gould, S. 1980. Caring groups and selfish genes. In *The panda's thumb*. New York: Norton.

Gregory, M., Silvers, A. & Sutch, D. 1978. *Sociobiology and human nature*. San Francisco: Jossey-Bass.

Grene, M. 1974. *The understanding of nature*. Boston: Reidel.

Grusec J. 1981. Socialization processes and the development of altruism. In *Altruism and helping behavior: Social, personality and developmental perspectives*. Edited by J. Rushton & S. Sorrentino, 65-90. Hillsdale, NJ: Erlbaum.

Guisinger, S. & Blatt, S. 1994. Individuality and relatedness: Evolution of a fundamental dialectic. *American Psychologist*. 45: 104-111.

Halevy, E. 1966. *The growth of philosophical radicalism*. Translated by M. Morris. Boston: The Beacon Press.

Hamilton, W. 1964a. The genetical evolution of social behavior. I. *Journal of Theoretical Biology* 7: 1-16.

———. 1964b The genetical evolution of social behavior. II. *Journal of Theoretical Biology* 7: 17- 32.

Hare, R. 1970. *Psychopathy, theory and research*. New York: Wiley.

———. 1981. *Moral thinking*. Oxford: Oxford University Press.

Harriman, P. 1974. *Handbook of psychological terms*. Totowa NJ: Littlefield.

Herzberg, A. ed. 1962. *Judaism*. New York: Brazillier.

Ho, M. 1984. Environment, heredity and development. In *Beyond neo-Darwinism. An introduction to the new evolutionary paradigm*. Edited by M. Ho & P. Saunders, 268-289. London: Academic Press.

Hobbes. 1962. *Leviathan*. Edited by M. Oakeshott. New York: Macmillan.

Hoffman, R. 1981. The development of empathy. In *Altruism and helping behavior: Social, personality, and developmental perspectives*. Edited by J. Rushton & R. Sorrentino, 41-63. Hillsdale, NJ: Erlbaum.

Holmes, W. & Sherman, P. 1983. Kin recognition in animals. *American Scientist*. 71: 46-55.

Hospers, J. 1966. What is explanation? In *Reason and responsibility*. Edited by J. Feinberg, 181-191. Belmont, CA: Dickenson Publishing Company.

Hull, D. 1988. Interactors and vehicles. In *The role of behavior in evolution*. Edited by H. Plotkin, 19-50. Cambridge: Massachusetts Institute of Technology Press.

_____. 1989. *The metaphysics of evolution*. New York: State University of New York Press.
Hume, D. 1938. Of liberty and necessity. In *The Harvard classics:* Vol. 37. Edited by C. Eliot, 351-370. New York: Collier.
_____. 1956. *An inquiry concerning the principles of morals*. New York: Bobbs, Merrill.
Huxley, L. ed. 1900. *Life and letters of Thomas Henry Huxley*. London: Macmillan.
Ingold, T. 1986. *Evolution and social life*. London: Cambridge University Press.
Jaworski, L. 1976. Honesty and professional ethics: Focus on law. In *The ethical basis of economic freedom*. Edited by J. Hill, 175-195. Chapel Hill, NC: America's Viewpoint, Inc.
Kauffman, S. 1993. *The origins of order: Self-organization and selection in evolution*. New York: Oxford University Press.
Kaye, H. 1986. *The social meaning of modern biology*. New Haven: Yale University Press.
King, J. 1980. The genetics of sociobiology. In *Sociobiology examined*. Edited by A. Montague, 82-107. London: Oxford University Press.
Koestler, A. 1967. *The ghost in the machine*. London: Hutcheson.
_____. 1978. *Janus: A summing up*. New York: Random House.
Kohlberg, L. 1971. From is to ought: How to commit the naturalistic fallacy and get away with it in the study of moral development. In *Cognitive Development and Epistemology*. Edited by T. Mischel, 127-136. New York: Academic Press.
_____. 1987. *Child psychology and childhood education*. New York: Longman.
Kohlberg, L. & Mayer R. 1972. Development as the aim of education. *Harvard Educational Review* 42: 478-482.
Kohn, H. 1966. *Political ideologies of the twentieth century*. Third Edition. New York: Harper.
Kowalski, G. 1980. In *Morality as a biological phenomenon: The presuppositions of sociobiology research*. Edited by G. Stent, 231-252. Revised edition. Berkeley: University of California Press.
Krebs, D. 1987. The challenge of altruism in biology and psychology. In *Sociobiology and psychology: Ideas, issues, and application*. Edited by C. Crawford, M. Smith & D. Krebs, 81-118. Hillsdale, NJ: Erlbaum Associates.

Kummer, H. 1980. Analogs of morality among nonhuman primates. In *Morality as a biological phenomenon:The presuppositions of sociobiology research.* Edited bv G. Stent, 31-47. Revised edition. Berkeley: University of California Press.

Kupfersmid, J. & Wonderly, D. *An author's guide to publishing better articles in better journals in the behavioral sciences.* Brandon VT. Clinical Psychology Publishing Co., Inc.

Laird, J. 1970. *An enquiry into moral notions.* New York: AMS Press.

Latane, B. & Rodin, J. 1969. A lady in distress: Inhibiting effects of friends and strangers on bystander intervention. *Journal of Experimental Social Psychology.* 5: 189-202.

Lawrence, E. 1989. *A guide to modern biology; Genetics, cells, and systems.* New York: Wiley.

Leakey, R. 1992. *Origins reconsidered: In search of what makes us human.* New York: Doubleday.

Lecky, W. 1955. *History of European morals from Augustus to Charlemagne.* Vol 1. pp. 178-181. New York: Braziller.

Levine, F. 1975. *Theoretical readings in motivation.* Chicago: Rand McNally.

Levins, R. & Lewontin, R. 1989. *The dialectical biologist.* Cambridge: Harvard University Press.

Li, W. & Grauer, D. 1991. *Fundamentals of molecular evolution.* Sunderland, MA: Sinauer.

Loomis, W. 1988. *Four billion years.* Sunderland, MA: Sinauer.

Lumsden, C. & Wilson, E. 1981. *Genes, mind, and culture: The coevolutionary process.* Cambridge: Harvard University Press.

MacDonald, K. 1988. Sociobiology and the cognitive-developmental tradition in moral development research. In *Sociobiological perspectives on human development.* Edited by K. MacDonald, 140-167. New York: Springer-Verlag.

McGrory, M. 1994. Whitewater to dilute GOP's own S & L mud. Jacksonville, FL: *The Florida Times Union,* 6 June.

McShea, R. 1990. *Morality and human nature.* Temple, PA: Temple University Press.

Mandelbaum, M., Gramlich F., Anderson R. & Schneewind J. eds. 1967. *Philosophic problems.* New York: Macmillan.

Markl, H. 1980. In *Morality as a biological phenomenon: The presuppositions of sociobiology research.* Edited bv G. Stent, 209-230. Revised edition. Berkeley: University of California Press.

Martin, E. ed. 1990. *A concise dictionary of biology.* Oxford: Market House Books.

Marx, M. & Hillix, W. 1979. *Systems and theories in psychology.* Third Edition. New York: McGraw Hill.

Masters, R. 1981. The value and limits of sociobiology. In *Sociobiology and human politics.* Edited by E. White, pp. 135-165. Lexington, MA: Lexington Books.

_____. 1989. *The nature of politics.* New Haven CT: Yale University Press.

Mayo, O. 1983. *Natural selection and its constraints.* New York: Academic Press.

Mayr, E. 1982. *The growth of biological thought.* Cambridge: Harvard University Press.

_____. 1988. *Toward a new philosophy of biology.* Cambridge: Belknap.

Menninger, K. 1938. *Man against himself.* New York: Harcourt, Brace & World.

Metchnikoff, E. 1977. *The nature of man.* Translated by P. Mitchell. New York: Arno Press.

Midgley, M. 1980. Rival fatalisms: The hollowness of the sociobiology debate. In *Sociobiology examined.* Edited by A. Montague, 15-38. London: Oxford University Press.

Mill, J. Of liberty and necessity. In *A modern introduction to philosophy.* Ed. by P. Edwards, & A. Pap, 44-50. New York: The Free Press

Monod, J. 1971. *Chance and necessity: An essay on the natural philosophy of modern biology.* Translated by A. Wainhouse. New York: Knopf.

Montague, A. ed. 1980. *Sociobiology examined.* London: Oxford University Press.

Moore, G. 1965. The indefinability of good. In *A modern introduction to philosophy.* Edited by P. Edwards & A. Pap, 321-327. New York: Collier-Macmillan.

Mussen, P., Rutherford, E., Harris, S. & Keasey, C. 1970. Honesty and altruism among preadolescents. *Developmental Psychology*, Vol. 3. 169-194.

Nagel, T. 1980. Ethics as an autonomous theoretical subject. In *Morality as a biological phenomenon: the presuppositions of sociobiology research.* Edited by G. Stent, 198-205. Revised edition. Berkeley: University of California Press.

Nowak M., May, R. & Sigmund, K. 1995. The arithmetic of mutual help. *Scientific American* Vol. 272 (6): 76-81.

Oldscamp, P. 1970. *The moral philosophy of George Berkeley*. The Hague: Nijhoff.

Oliwenstein, L. Death and the microbe. In *Discover*. Sept 1995: pp. 99-100.

O'Rourke, P. 1991. *A Parliament of Whores*. New York: The Atlantic Monthly Press.

Ortner, D. 1983. Biocultural interaction in human adaptation. In *How humans adapt. A biocultural odyssey*. Edited by D. Ortner, 127-162. Washington, D. C.: Smithsonian Institution Press.

Parrot, M. 1976. Honesty and professional ethics: Focus on medicine. In *The ethical basis of economic freedom*. Edited by J. Hill, 161-172. Chapel Hill, NC: America's Viewpoint, Inc.

Peacock, J. 1976. Ethics, economics and society in evolutionary perspective. In *The ethical basis of economic freedom*. Edited by J. Hill, 21-37. Chapel Hill, NC: America's Viewpoint, Inc.

Plotkin, H. 1988. Behavior and evolution. In *The role of behavior in evolution*. Edited by H. Plotkin, 1-18. Cambridge: Massachusetts Institute of Technology Press.

Porter, R. 1987. Kin recognition: Functions and mediating mechanisms. In *Sociobiology and psychology: Ideas, issues, and application*. Edited by C. Crawford, M. Smith & D. Krebs, 175-203. Hillsdale NJ: Erlbaum Associates.

Portman, A. 1949. *Probleme des lebens*. Basel: Reinhart.

Prichard, H. 1975. Duty and interest. In *Reason and responsibility*. Edited by J. Feinberg, 386-394. Belmont, CA: Dickenson.

Rachels, J. 1991. *Created from animals: The moral implications of Darwinism*. London: Oxford University Press.

Reid, T. 1965. The moral faculty and the principles of morals. In *A modern introduction to philosophy*. Edited by P. Edwards & A. Pap, 288-296. New York: Collier-Macmillan.

Rheingold, H. & Hay, D. 1980. Prosocial behavior of the very young. In *Morality as a biological phenomenon: The presuppositions of sociobiology research*. Edited by G. Stent, 93-106. Revised edition. Berkeley: University of California Press.

Ridley, M. 1985. *The problems of evolution*. New York: Oxford University Press.

Riviere, J. ed. & trans. 1959. *Sigmund Freud: collected papers*. Vol. 4. New York: Basic Books.

Rosen, R. 1991. *Life itself: A comprehensive inquiry into the nature, origin, and fabrication of life*. New York: Columbia University Press.

Rosenberg, A. 1984. *The structure of biological science.* Cambridge: Cambridge University Press.
Roszak, T. 1972. *Where the wasteland ends.* New York: Doubleday.
Royce, J. 1892. *The spirit of modern philosophy.* Boston: Houghton Mifflin.
Runes, D. ed. 1962. *Dictionary of philosophy.* Totowa, NJ: Littlefield,
Ruse, M. 1988. *Philosophy of biology today.* Albany: State University of New York Press.
Russell, B. 1945. *A history of western philosophy.* New York: Simon & Schuster.
Ryle, G. 1949. *The concept of mind.* New York: Hutchinson's University Library.
Schwartz, B. 1986. *The battle for human nature.* New York: Norton & Company.
Scott, A. 1986. *The creation of life.* New York: Basil Blackwell Inc.
Segal, N. 1988. Cooperation, competition, and altruism in human twinships. In *Sociobiological perspectives on human development.* Edited by K. MacDonald, 168-206. New York: Springer-Verlag.
Sharp, F. 1928. *Ethics.* New York: Appleton-Century.
Shapiro, R. 1986. *Origins.* New York: Summit Books.
Simon, H. 1990. A mechanism for social selection and successful altruism. *Science.* 250: 1665-1668.
Sinha, C. 1984. A socio-naturalistic approach to human development. In *Beyond neo-Darwinism. An introduction to the new evolutionary paradigm.* Edited by M. Ho & P. Saunders, 331-362. London: Academic Press.
Skinner, B. 1971. *Beyond freedom and dignity.* NY: Knopf.
Smith, J. 1980. The concept of sociobiology. In *Morality as a biological phenomenon: The presuppositions of sociobiology research.* Edited bv G. Stent, 21-29. Revised edition. Berkeley: University of California Press.
_____. 1986. *The problems of biology.* New York: Oxford University Press.
_____. 1989a. *Evolutionary genetics.* London: Oxford University Press.
_____. 1989b. The debate continues. In *From Gaia to selfish genes.* Edited by C. Barlow, 238-239. London: Massachusetts Institute of Technology Press.
Smith, M. 1988. Research in developmental sociobiology. In *Sociobiological perspectives on human development.* Edited by K. MacDonald, 271-292. New York: Springer-Verlag.

Smythies, J. 1965. The representative theory of perception. In *Brain and mind: Modern concepts of the nature of mind*. Edited by J. Smythies, 241-257. London: Routledge & Kegan Paul.

Solomon, R. 1980. Group three. In *Morality as a biological phenomenon: The presuppositions of sociobiology research*. Edited by G. Stent, 253-274. Revised edition. Berkeley: University of California Press.

Spinoza, B. 1963. *Short treatise on God, man and his well being*. London: Routledge & Kegan Paul.

Sprague, E. 1967. Moral sense. In *The encyclopedia of philosophy*. Edited by P. Edwards Vol. 5: 385-387. New York: Collier MacMillan.

Stebbins, G. 1982. *Darwin to DNA: Molecules to humanity*. New York: Freeman.

Stent, G., ed. 1980. *Morality as a biological phenomenon: The presuppositions of sociobiology research*. Revised edition. Berkeley: University of California Press.

Stevenson, E., ed. 1959. *A Henry Adams reader*. Garden City New York: Doubleday.

Stroud, B. 1977. *Hume*. London: Routledge & Kegan Paul.

Taylor, P. 1964. *The moral judgment*. Princeton: Prentice Hall.

The Random House Dictionary of the English Language, 1987. Edited by S. Flexner & L Hauck. New York: Random House.

Thilly, F. 1936. *A history of philosophy*. New York: Holt.

Trigg, R. 1983. *The shaping of man: Philosophical aspects of sociobiology*. New York: Schocken.

Trivers, R. 1971. The evolution of reciprocal altruism. *Quarterly Review of Biology*. 46: 35-57.

Turiel, E. 1980. The development of moral concepts. In *Morality as a biological phenomenon: The presuppositions of sociobiology research*. Edited by G. Stent, 109-123. Revised edition. Berkeley: University of California Press.

Uyenoyama, M. & Feldman, M. 1980. Theories of kin and group selection: A population genetics perspective. *Theoretical population biology* 17: 380-414.

Waddington. C. 1971. *Biology, purpose, and ethics*. Worcester MA: Clark University Press.

Wald, G. 1979. The origin of life. In *Life: Origin and evolution*. Readings from Scientific American, 47-56. Edited by C. Fulsome. San Franscisco: Freeman.

Wallace, A. 1987. *Theories of life*. New York: Penguin.

Warnock, G. 1971. *The object of morality*. London: Methuen & Co. Ltd.
Weiss, E. 1960. *The structure and dynamics of the human mind*. London: Grune & Stratton.
White, R. 1959. Motivation reconsidered: The concept of competence. *Psychological Review* 66: 297-329.
Wicken, J. 1987. *Evolution, thermodynamics, and information*. New York: Oxford University Press.
Williams, B. 1980. Conclusion. In *Morality as a biological phenomenon: The presuppositions of sociobiology research*. Edited by G. Stent, 275-287. Revised edition. Berkeley: University of California Press.
Williams, G. 1984. Group selection. In *Genes, organisms, populations*. Edited by R. Brandon & R. Burian, 52-68. Cambridge: Massachusetts Institute of Technology Press.
_____. 1989. A sociobiological expansion of evolution and ethics. In *Evolution and ethics*. Edited by J. Paradis & G. Williams, 179-214. Princeton: Princeton University Press.
Wills, C. 1989. *The wisdom of the genes*. New York: Basic Books.
Wilson, D. 1980. *The natural selection of populations and communities*. Menlo Park CA: Cummings.
Wilson, E. 1975. *Sociobiology*. Cambridge, MA: Belknap.
Wilson, J. 1987. *The moral sense*. Totowa, NJ: Barnes & Noble.
Wolff, K. 1974. *Trying sociology*. New York: Wiley.
Wolff, P. 1980. The biology of morals from a biological perspective. In *Morality as a biological phenomenon: The presuppositions of sociobiology research*. Edited by G. Stent, 83-92. Revised edition. Berkeley: University of California Press.
Wonderly, D. 1991. *Motivation, behavior, and emotional health*. Lanham, MD: University Press of America.
Wonderly, D., & Kupfersmid, J. 1978. A test of the cognitive-developmental disequilibrium hypothesis in moral development. *The Journal of Psychology*, 100: 297-304.
_____. The relationship between moral judgment and selected characteristics of mental health. *Character Potential* 9(2): 111-116.
_____. 1980. Promoting postconventional morality: The adequacy of Kohlberg's aim. *Adolescence* XV(59): 609-631.
Yarrow, M., Scott, P. & Waxler, C. 1973. Learning concern for others. *Developmental psychology*, 8: 240-260.
Zukav, G. 1979. *The dancing Wu Li masters*. New York: Bantam.

Index

A
adaptation, 61-63
affect, 97, 98
Albert, E., 3
Alland, A., 14
altruism, 11-13
    and ethics, 7
    and morality, 114-17
    and reason, 114-117
Ardrey, R., 14
Aresty, E., 186
Aristotle, 3, 187

B
Baier, K., 5
BAP, 107-110
Batson, C., 37
behavior, 102, 107-111
    and belief, 98
    critical characteristics, 111-114
    incident, 109
    instrumental, 109, 119
    intrinsic, 109, 110, 119
    outcome orientation, 111, 112
    pattern, 110
    total sequence, 110
belief, 98
    and behavior, 98
    defined, 98, 118
Berman, H., 188
Bernal, J., 65
Blatt, S., 74, 184
Boden, M., 9
Bohm, D., 63
Brandon, R., 12, 44, 46
Burian, R., 12, 44, 46
Butler, J., 32

C
Campbell, R., 10
Caporeal, L., 15
Cattell, R., 18
cognition, 97, 98
communication, 74
costs, 107-110, 113
creationism, 39-41
Crick, F., 28,
culturism 32-37

D

Dawkins, R., *xvi, xvii,*11, 30, 44, 47, 61, 74, 76, 134, 135, 142, 145, 148-150, 151, 152, 153, 154, 156, 160, 163,164, 166, 167, 169, 170, 187, 211
Day, W., 59
decisions, 109, 110
deliberation, 109, 110
desire, 96, 98-102
    assertive, 99
    protective, 99
    transcendent, 100, 196-200
    as instinctive knowledge, 102, 118
Dobzhansky, T., 45, 58, 59, 183
Dyson, F., 28

E

emergence of morality, 13-17
emotion, 105, 106
Ehrlich, P., 187
ethics, 2-8
    and the law, 190-192
    legal, 192, 193
    medical, 193, 194
evolution, 58-60

F

Feder, D., *xx, xxi, xxiv*
Ford, C., 190
Flew, A., 2
Fox, R., 16, 35, 47, 48, 57
Fox, R. F., 58, 59
Fox, R. L., 16
Franklin, B., 109
Freedman, D., 59
Freud, A., 18
Freud, S., *vi, vii,* 18
Fromm, E., 18

G

genes, 72-76
    altruistic, 201-204
    as innate knowledge, 73
    defined, 74
Ghiselin, M., 29, 41, 154
Grauer, D., 44, 60, 137
Grene, M., 9
Guisinger, S., *xxi,* 107, 108, 188, 204, 205

H

Hamilton, W., *xvi,* 28, 30, 137, 138, 140-142, 145-148, 157, 158, 162, 210
Hare, R., 5
Hill, J., 192
Ho, M., 59
Hobbes. T., 30, 136, 191
Hoffman, R., 41, 42
holarchy, 70, 71
holon, 63-67
    directional, 65-67
Holm, R., 187
Hull, D., 37, 44, 74
Hume, D., 6, 7, 31, 33, 103, 117-120, 159, 191, 192, 199

I

identification, 107
information, 72
instance, 65
instinct, 102
    knowledge as, 102
    behavior as, 102
intuitionism, 6-8

J

Jaworski, L., 193
Jung, K., 32

Kant, I., 4
Kaufman, S., 8
Kaye, H., 29, 30, 33
knowledge, 71, 72
    defined, 71
    innate, 71
    learned, 71
Koestler, A., 64, 70, 103
Kohlberg, L., 3, 38, 39
Kohn, H., 191
Kowalski, G., 19, 116
Krebs, D., 41, 165, 166
Kummer, H., 36, 169, 186
Kupfersmid, J., 39

L
Laird, J., 7, 19
Latane, B. 41, 42,
Leakey, R., 5, 16, 61
Lecky, W., 14
Levins, R., 58
Lewontin, R., 33, 58
Li, W., 44, 59, 60, 133
life defined, 66
Locke, J., 6, 7
Lumsden, C., 47

M
Mandelbaum, M., 14
Markl, H., 6, 35
Masters, R., 48
Mayo, O., 61
Mayr, E., 30, 36, 44, 59-62, 132,
    151, 166, 180, 183
Menninger, K., 18
Metchnikoff, E., 18
Mill, J., 4, 6
mind, 95-97
    defined, 96
Monod, J., 33

Montague, A., 34
Moore. G., 5, 7
morality, 1-4, 103, 104
Mussen, P., 43

N
Nagel, T., 35
need, 102, 103
    defined, 102
Nietszche, F., 30

P
Parrot, M., 189, 190
part, 65
Peacock, J., 45
perception, 96
Plato, 6, 186
Portman, A., 74
Prichard, H., 5
provincialism, 8-10
psychological interpretations, 37-39
psychosocial models, 41-43

R
Rachels, J., 182
reason, 103
Ridley, M., 61
Reid, T., 33,
Rodin, J., 41, 42
Rosen, R., 8, 60
Rosenberg, A., 9, 179
Royce, J., 31
Rune, D., 2
Ruse, M., 1, 2, 6, 7, 14, 45, 46, 61
Russell, B., 93

S
Schwartz, B., 35
Scott A., 36
Sharp, F., 5

Simon, H., 43
Sinha, C., 59
Smith, J., 33, 133, 140, 141, 157, 159, 160
Socrates, 1
Solomon, R., 11, 112
Spinoza, B., 1, 3
Stebbins, G., 15, 36, 45, 61, 129, 137
Stent, G., 36, 37
Stroud, B., 98
systems, 67-70
    defined, 67
    inorganic, 68
    organic, 68

T

Taylor, P., 19
The good and the natural, 10-11
The self, 31, 71, 72
    and altruism, 75
    sensory systems, 93-95
The unit of selection, 44-48

theological interpretations, 39-41
Trivers, R., 30, 128-140, 145-147, 159
Turiel, E., 19

W

Waddington, C., 2, 8, 29-32 35,
want/cost ratio, 114
    and behavior, 114
Warnock, G., 5, 14, 182, 183, 201
Williams, G., 45, 46, 75, 131, 136, 142, 166, 206
whole, 65
Williams, B., 34
willing, 101, 110, 111
Wills, C., 61
Wilson, D., 201
Wilson, E., 27, 29, 34, 45, 130, 151, 152, 162
Wolff, K., 8
Wolff, P., 32, 38, 40
Wonderly, D., 39